Preston Albert Lambert

Analytic Geometry

For Technical Schools and Colleges

Preston Albert Lambert

Analytic Geometry
For Technical Schools and Colleges

ISBN/EAN: 9783744646093

Printed in Europe, USA, Canada, Australia, Japan

Cover: Foto ©berggeist007 / pixelio.de

More available books at **www.hansebooks.com**

ANALYTIC GEOMETRY

FOR

TECHNICAL SCHOOLS AND COLLEGES

BY

P. A. LAMBERT, M.A.
INSTRUCTOR IN MATHEMATICS, LEHIGH UNIVERSITY

New York
THE MACMILLAN COMPANY
LONDON: MACMILLAN & CO., Ltd.
1897

All rights reserved

COPYRIGHT, 1897,
BY THE MACMILLAN COMPANY.

Norwood Press
J. S. Cushing & Co.   Berwick & Smith
Norwood Mass. U.S.A.

PREFACE

The object of this text-book is to furnish a natural but thorough introduction to the principles and applications of Analytic Geometry for students who have a fair knowledge of Elementary Geometry, Algebra, and Trigonometry.

The presentation is descriptive rather than formal. The numerous problems are mainly numerical, and are intended to give familiarity with the method of Analytic Geometry, rather than to test the student's ingenuity in guessing riddles. Answers are not given, as it is thought better that the numerical results should be verified by actual measurement of figures carefully drawn on cross-section paper.

Attention is called to the applications of Analytic Geometry in other branches of Mathematics and Physics. The important engineering curves are thoroughly discussed. This is calculated to increase the interest of the student, aroused by the beautiful application the Analytic Geometry makes of his knowledge of Algebra. The historical notes are intended to combat the notion that a mathematical system in all its completeness issues Minerva-like from the brain of an individual.

P. A. LAMBERT.

TABLE OF CONTENTS

ANALYTIC GEOMETRY OF TWO DIMENSIONS

CHAPTER I

RECTANGULAR COORDINATES

ARTICLE		PAGE
1.	Introduction	1
2.	Coordinates	1
3.	The Point in a Straight Line	2
4.	The Point in a Plane	3
5.	Distance between Two Points	7
6.	Systems of Points in the Plane	8

CHAPTER II

EQUATIONS OF GEOMETRIC FIGURES

7.	The Straight Line	13
8.	The Circle	15
9.	The Conic Sections	15
10.	The Ellipse	18
11.	The Hyperbola	21
12.	The Parabola	24

CHAPTER III

PLOTTING OF ALGEBRAIC EQUATIONS

13.	General Theory	28
14.	Locus of First Degree Equation	29
15.	Straight Line through a Point	30
16.	Tangents	31

ARTICLE	PAGE
17. Points of Discontinuity	33
18. Asymptotes	34
19. Maximum and Minimum Ordinates	36
20. Points of Inflection	37
21. Diametric Method of Plotting Equations	39
22. Summary of Properties of Loci	39

CHAPTER IV

PLOTTING OF TRANSCENDENTAL EQUATIONS

23. Elementary Transcendental Functions	45
24. Exponential and Logarithmic Functions	45
25. Circular and Inverse Circular Functions	47
26. Cycloids	54
27. Prolate and Curtate Cycloids	57
28. Epicycloids and Hypocycloids	58
29. Involute of Circle	59

CHAPTER V

TRANSFORMATION OF COORDINATES

30. Transformation to Parallel Axes	60
31. From Rectangular Axes to Rectangular	61
32. Oblique Axes	62
33. From Rectangular Axes to Oblique	63
34. General Transformation	64
35. The Problem of Transformation	65

CHAPTER VI

POLAR COORDINATES

36. Polar Coordinates of a Point	70
37. Polar Equations of Geometric Figures	71
38. Polar Equation of Straight Line	71
39. Polar Equation of Circle	72
40. Polar Equations of the Conic Sections	73
41. Plotting of Polar Equations	76
42. Transformation from Rectangular to Polar Coordinates	79

CONTENTS

CHAPTER VII
PROPERTIES OF THE STRAIGHT LINE

ARTICLE		PAGE
43.	Equations of the Straight Line	81
44.	Angle between Two Lines	84
45.	Distance from a Point to a Line	85
46.	Equations of Bisectors of Angles	86
47.	Lines through Intersection of Given Lines	87
48.	Three Points in a Straight Line	88
49.	Three Lines through a Point	89
50.	Tangent to Curve of Second Order	91

CHAPTER VIII
PROPERTIES OF THE CIRCLE

51.	Equation of the Circle	93
52.	Common Chord of Two Circles	94
53.	Power of a Point	95
54.	Coaxal Systems	97
55.	Orthogonal Systems	98
56.	Tangents to Circles	101
57.	Poles and Polars	102
58.	Reciprocal Figures	104
59.	Inversion	106

CHAPTER IX
PROPERTIES OF THE CONIC SECTIONS

60.	General Equation	111
61.	Tangents and Normals	113
62.	Conjugate Diameters	119
63.	Supplementary Chords	122
64.	Parameters	124
65.	The Elliptic Compass	126
66.	Area of the Ellipse	127
67.	Eccentric Angle of Ellipse	128
68.	Eccentric Angle of the Hyperbola	130

CHAPTER X

Second Degree Equation

ARTICLE		PAGE
69.	Locus of Second Degree Equation	133
70.	Second Degree Equation in Oblique Coordinates	138
71.	Conic Section through Five Points	141
72.	Conic Sections Tangent to Given Lines	142
73.	Similar Conic Sections	144
74.	Confocal Conic Sections	146

CHAPTER XI

Line Coordinates

75.	Coordinates of a Straight Line	149
76.	Line Equations of the Conic Sections	151
77.	Cross-ratio of Four Points	151
78.	Second Degree Line Equations	152
79.	Cross-ratio of a Pencil of Four Rays	153
80.	Construction of Projective Ranges and Pencils	155
81.	Conic Section through Five Points	157

CHAPTER XII

Analytic Geometry of the Complex Variable

82.	Graphic Representation of the Complex Variable	160
83.	Arithmetic Operations applied to Vectors	162
84.	Algebraic Functions of the Complex Variable	165
85.	Generalized Transcendental Functions	168

ANALYTIC GEOMETRY OF THREE DIMENSIONS

CHAPTER XIII

Point, Line, and Plane in Space

86.	Rectilinear Space Coordinates	171
87.	Polar Space Coordinates	173
88.	Distance between Two Points	174

ARTICLE		PAGE
89.	Equations of Lines in Space	176
90.	Equations of the Straight Line	177
91.	Angle between Two Straight Lines	179
92.	The Plane	182
93.	Distance from a Point to a Plane	184
94.	Angle between Two Planes	186

CHAPTER XIV

CURVED SURFACES

95.	Cylindrical Surfaces	190
96.	Conical Surfaces	191
97.	Surfaces of Revolution	193
98.	The Ellipsoid	194
99.	The Hyperboloids	195
100.	The Paraboloids	196
101.	The Conoid	197
102.	Equations in Three Variables	198

CHAPTER XV

SECOND DEGREE EQUATION IN THREE VARIABLES

103.	Transformation of Coordinates	200
104.	Plane Section of a Quadric	201
105.	Center of Quadric	203
106.	Tangent Plane to Quadric	203
107.	Reduction of General Equation of Quadric	205
108.	Surfaces of the First Class	208
109.	Surfaces of the Second Class	210
110.	Surfaces of the Third Class	211
111.	Quadrics as Ruled Surfaces	212
112.	Asymptotic Surfaces	213
113.	Orthogonal Systems of Quadrics	214

ANALYTIC GEOMETRY

CHAPTER I

RECTANGULAR COORDINATES

Art. 1. — Introduction

The object of analytic* geometry is the study of geometric figures by the processes of algebraic analysis.

The three fundamental problems of analytic geometry are:

To find the equation of a geometric figure or the equations of its several parts from its geometric definition.

To construct the geometric figure represented by a given equation.

To find the relations existing between the geometric properties of figures and the analytic properties of equations.

Art. 2. — Coordinates

Any scheme by means of which a geometric figure may be represented by an equation is called a system of coordinates.

---

* The reasoning of pure geometry, the geometry of Euclid, is mainly synthetic, that is, starting from something known we pass from consequence to consequence until something new results. The reasoning of algebra is analytic, that is, assuming what is to be demonstrated we pass from consequence to consequence until the relation between the unknown and the known is found. The term "analytic geometry" is therefore equivalent to algebraic geometry. The application of algebra to the determination of the properties of geometric figures was invented by Descartes (1596-1650), a French philosopher, and published in Leyden in 1637.

The coordinates of a point are the quantities which determine the position of the point.

Along the line of a railroad the position of a station is determined by its distance and direction from a fixed station; on our maps the position of a town is determined by its latitude and longitude, the distances and directions of the town from two fixed lines of the map; the position of a point in a survey is determined by its distance and bearing from a fixed station.

On these different methods of determining the position of a point are based different systems of coordinates.

Art. 3. — The Point in a Straight Line

On a straight line a single quantity or coordinate is sufficient to determine the position of a point. Let 0 be a fixed point

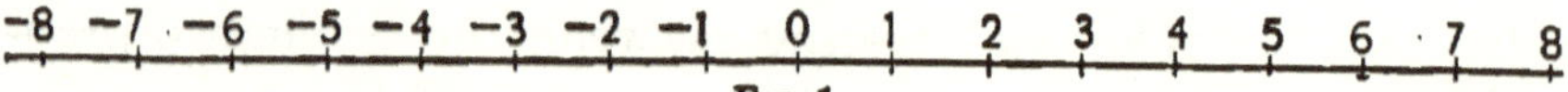

$$-8\ -7\ -6\ -5\ -4\ -3\ -2\ -1\ \ 0\ \ 1\ \ 2\ \ 3\ \ 4\ \ 5\ \ 6\ \ 7\ \ 8$$

Fig. 1.

in the line; adopt some length, such as 01, as the linear unit; call distances measured from 0 towards the right positive, distances measured from 0 towards the left negative. Let a point of the line be represented by the number which expresses its distance and direction from the fixed point 0. Then to every real number, positive or negative, rational or irrational, there corresponds a definite point in the straight line, and to every point in the line there corresponds a definite real number. This fact is expressed by saying that there is a "one-to-one correspondence" between the points of the line and real numbers.

The algebra of a single real variable finds a geometric interpretation in the straight line. Denoting by x the distance and direction of a point in the straight line from 0, that is letting x denote the coordinate of the point, the equation $x^2 - 2x - 8 = 0$ locates the two points (4), (−2), in the straight line.

RECTANGULAR COORDINATES

Problems. — 1. Locate in the straight line the points 3; -2; $1\frac{1}{2}$; -2.5; -5; $\frac{3}{5}$.

2. Locate $\sqrt{5}$; $-\sqrt{8}$; $\sqrt{10}$; $\sqrt{7}$.

SUGGESTION. — The numerical value of $\sqrt{5}$ can be found only approximately. The hypotenuse of a right triangle whose two sides about the right angle are 2 and 1, represents $\sqrt{5}$ exactly.

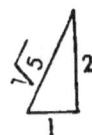

FIG. 2.

3. Find the point midway between x_1 and x_2.

4. Find the point dividing the line from x_1 to x_2 internally into segments whose ratio is r.

5. Find the point dividing the line from x_1 to x_2 externally into segments whose ratio is r.

6. Locate the roots of $x^2 + 2x - 8 = 0$.

7. Locate the roots of $x^2 - 4x - 4 = 0$.

8. Locate the roots of $x^3 - 6x^2 + 11x - 6 = 0$.

9. Find the points dividing into three equal parts the line from 2 to 14.

10. Find the points dividing into three equal parts the line from x_1 to x_2.

11. Find the point dividing a line 8 feet long internally into segments in the ratio $3:4$.

12. A uniform bar 10 feet long has a weight of 15 pounds at one end, of 25 pounds at the other end. Find the point of support for equilibrium.

ART. 4. — THE POINT IN A PLANE

To determine the position of a point in a plane, assume two straight lines at right angles to each other to be fixed in the plane. These lines are called the one the X-axis, the other the Y-axis. The distance from a point in the plane to either axis is measured on a line parallel to the other axis; the direction of the point from the axis is indicated by the algebraic sign prefixed to the number expressing the distance from the axis.

4   ANALYTIC GEOMETRY

Distances measured parallel to the X-axis to the right from the Y-axis are called positive; those measured to the left from the Y-axis are called negative. The distance and direction of a point from the Y-axis is called the abscissa of the point, and is denoted by x.

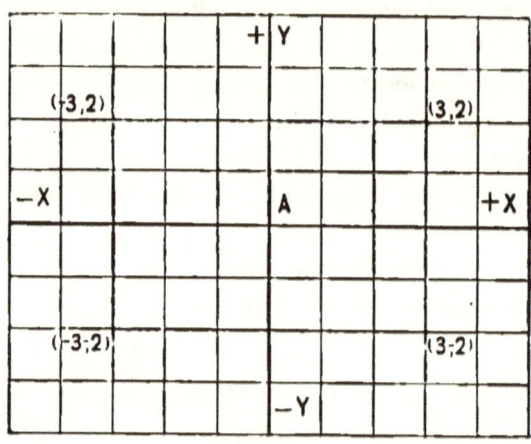

Fig. 3.

Distances measured parallel to the Y-axis upward from the X-axis are called positive; those measured downward from the X-axis are called negative. The distance and direction of a point from the X-axis is called the ordinate of the point, and is denoted by y.

The axes of reference cut the plane into four parts. Calling the part $+XA+Y$ the first angle, $+YA-X$ the second angle, $-XA-Y$ the third angle, $-YA+X$ the fourth angle, it is seen that in the first angle ordinate and abscissa are both positive; in the second angle the ordinate is positive, the abscissa negative; in the third angle ordinate and abscissa are both negative; in the fourth angle the ordinate is negative, the abscissa positive.

The abscissa of a point determines a straight line parallel to the Y-axis in which the point must lie. For, by elementary geometry, the locus of all points on one side of a straight line and equidistant from the straight line is a straight line parallel to the given line.

The ordinate of a point determines a straight line parallel to the X-axis in which the point must lie.

If both ordinate and abscissa of a point are known, the point must lie in each of two straight lines at right angles to each other, and must, therefore, be the intersection of these lines. Hence ordinate and abscissa together determine a single point in the plane.

Conversely, to a point in the plane there corresponds one ordinate and one abscissa. For through the point only one straight line parallel to the Y-axis can be drawn. This fact determines a single value for the abscissa of the point. Through the given point only one parallel to the X-axis can be drawn. This determines a single value for the ordinate of the point.

The abscissa and ordinate of a point as defined are together the rectangular * coordinates of the point. The point whose coordinates are x and y is spoken of as the point (x, y). There is a "one-to-one correspondence" between the symbol (x, y) and the points of the XY-plane.

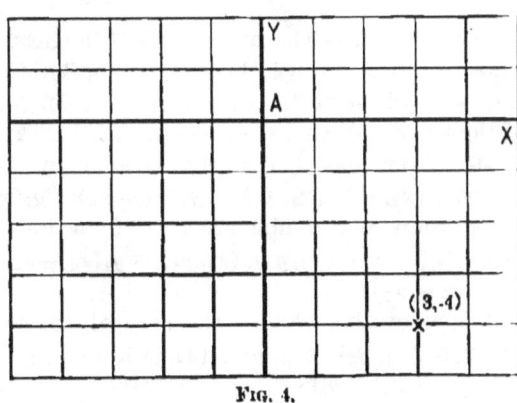

Fig. 4.

Problems. — 1. Locate the point $(3, -4)$.

Lay off 3 linear units on the X-axis to the right from the origin, and there is found the straight line parallel to the Y-axis, in which the point must lie. On this line lay off 4 linear units downward from its intersection with the X-axis, and the point $(3, -4)$ is located.

---

* This method of representing a point in a plane was invented by Descartes. Hence these coordinates are also called Cartesian coordinates.

2. Locate $(-3, 0)$; $(0, 4)$; $(1, -1)$; $(-1, -1)$; $(-7, 5)$; $(10, -7)$; $(15, 20)$.

3. Locate $(2\tfrac{1}{2}, 3)$; $(-1, 3\tfrac{1}{2})$; $(1\tfrac{1}{2}, -5\tfrac{1}{2})$; $(7.8, -4.5)$.

4. Locate $(\sqrt{2}, \sqrt{5})$; $(-\sqrt{8}, \sqrt{17})$; $(\sqrt{50}, \sqrt{75})$.

5. Construct the triangle whose vertices are $(4, 5)$, $(-2, 7)$, $(-3, -6)$.

6. Find the point midway between $(4, 7)$, $(6, 5)$.

7. Find the point midway between (x', y'), (x'', y'').

8. Find the area of the triangle whose vertices are $(0, 0)$, $(0, 8)$, $(9, 0)$.

9. Find the area of the triangle whose vertices are $(2, 1)$, $(5, 4)$, $(9, 2)$.

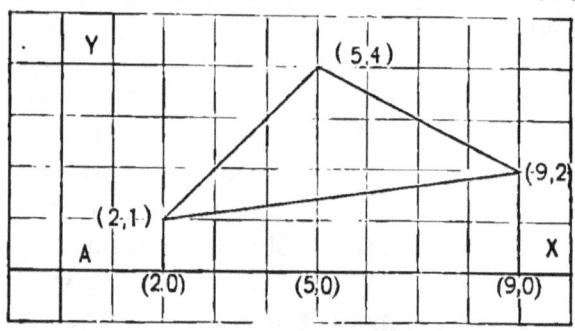

Fig. 5.

Suggestion.—The area of the triangle is the area of the trapezoid whose vertices are $(2, 1)$, $(2, 0)$, $(5, 4)$, $(5, 0)$, plus the area of the trapezoid whose vertices are $(5, 4)$, $(5, 0)$, $(9, 2)$, $(9, 0)$, minus the area of the trapezoid whose vertices are $(2, 1)$, $(2, 0)$, $(9, 2)$, $(9, 0)$.

10. Show that double the area of the triangle whose vertices are (x_1, y_1), (x_2, y_2), (x_3, y_3) is $y_1(x_3 - x_2) + y_2(x_1 - x_3) + y_3(x_2 - x_1)$.

11. Show that double the area of the quadrilateral whose vertices are (x_1, y_1), (x_2, y_2), (x_3, y_3), (x_4, y_4) is $y_1(x_4 - x_2) + y_2(x_1 - x_3) + y_3(x_2 - x_4) + y_4(x_3 - x_1)$.

12. Show that double the area of the pentagon whose vertices are (x_1, y_1), (x_2, y_2), (x_3, y_3), (x_4, y_4), (x_5, y_5) is $y_1(x_5 - x_2) + y_2(x_1 - x_3) + y_3(x_2 - x_4) + y_4(x_3 - x_5) + y_5(x_4 - x_1)$.

Notice that double the area of any polygon is the sum of the products of the ordinate of each vertex by the difference of the abscissas of the adjacent vertices, these differences being taken in the same direction, anti-clockwise, around the entire polygon.

13. Find the area of the triangle whose vertices are $(12, -5)$, $(-8, 7)$, $(10, 15)$.

14. Find the condition that (x, y) lie in the straight line through (x', y'), (x'', y'').

15. Show that the points $(1, 4)$, $(3, 2,)$, $(-3, 8)$ lie in a straight line.

16. The vertices of a pentagon are $(2, 3)$, $(-5, 8)$, $(11, -4)$, $(9, 12)$, $(14, 7)$. Plot the pentagon and find its area.

17. A piece of land is bounded by straight lines. From the survey the rectangular coordinates of the stations at the corners referred to a N.S. line and an E.W. line through station A are as follows, distances measured in chains:

A	0	0	D	22.85	17.19
B	14.30	-15.04	E	7.42	40.09
C	22.85	-4.18	F	-8.29	29.80

Plot the survey and find the area of the piece of land.

18. Find the point which divides the line from (x', y') to (x'', y'') internally into segments whose ratio is r.

19. Find the point which divides the line from (x', y') to (x'', y'') externally into segments whose ratio is r.

20. Locate the points $(2, -9)$, $(-6, 5)$, and also the points dividing the line joining them internally and externally in the ratio $2:3$.

21. Show that the points (x, y), $(x, -y)$ are symmetrical with respect to the X-axis.

22. Show that the points (x, y), $(-x, y)$ are symmetrical with respect to the Y-axis.

23. Show that the points (x, y), $(-x, -y)$ are symmetrical with respect to the origin.

Art. 5.— Distance between Two Points

The distance between the points (x', y'), (x'', y'') is the hypotenuse of the right triangle whose two sides about the right angle are $(x' - x'')$ and $(y' - y'')$. Hence

$$d = \sqrt{(x' - x'')^2 + (y' - y'')^2}.$$

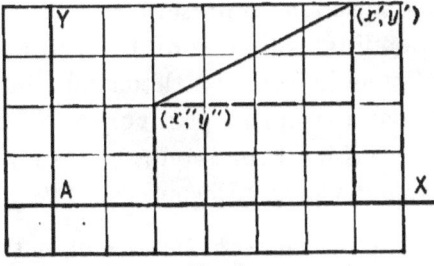

Fig. 6.

Problems.—1. Find distance between the points $(4, 2)$, $(7, 5)$; $(-3, 6)$, $(4, -9)$; $(0, 8)$, $(7, 0)$; $(15, -17)$, $(8, 2)$; $(-4, -7)$, $(-12, -19)$.

2. Derive formula for distance from (x', y') to the origin.

3. Find distance from origin to $(5, 9)$; $(7, -4)$; $(12, -15)$; $(-9, 14)$.

4. Find the lengths of the sides of the triangle whose vertices are $(-3, -2)$, $(7, 8)$, $(-5, 6)$.

5. The vertices of a triangle are $(0, 6)$, $(4, -5)$, $(-2, 8)$. Find the lengths of the medians.

6. Find the distance between the middle points of the diagonals of the quadrilateral whose vertices are $(2, 3)$, $(-4, 5)$, $(6, -3)$, $(0, 7)$.

7. Show that the points $(6, 6)$, $(1\frac{1}{2}, 15)$, $(-3, -12)$, $(-7\frac{1}{2}, -3)$ are the vertices of a parallelogram.

8. Find the center of the circle circumscribing the triangle whose vertices are $(2, 2)$, $(7, -3)$, $(2, -8)$.

9. Find the equation which expresses the condition that the point (x, y) is equidistant from $(4, -5)$, $(-3, 7)$.

10. Find the equation which expresses the condition that the distance from the point (x, y) to the point $(-3, 2)$ is 5.

11. Find the equation which locates the point (x, y) in the circumference of a circle whose radius is r, center (a, b).

Art. 6. — Systems of Points in the Plane

If any two quantities, which may be called x and y, are so related that for certain values of x, the corresponding values of y are known, the different pairs of corresponding values of x and y may be represented by points in the XY-plane.

Comparative statistics and experimental results can frequently be more concisely and more forcibly presented graphically than by tabulating numerical values. In the diagram the abscissas represent the years from 1878 to 1891, the corresponding ordinates of the full and dotted lines the production of steel in hundred thousand long tons in the United States and Great Britain respectively.* The diagram exhibits graphically the information contained in the adjacent table, condensed from "Mineral Resources," 1892. Observe that if the points are

---

* In the figure the linear unit on the X-axis is 5 times the linear unit on the Y-axis. It will be noticed that the essential feature of a system of coordinates, the "one-to-one correspondence" of the symbol (x, y) and the points of the XY-plane, is not disturbed by using different scales for ordinates and abscissas.

inaccurately located the diagram becomes not only worthless, but misleading.

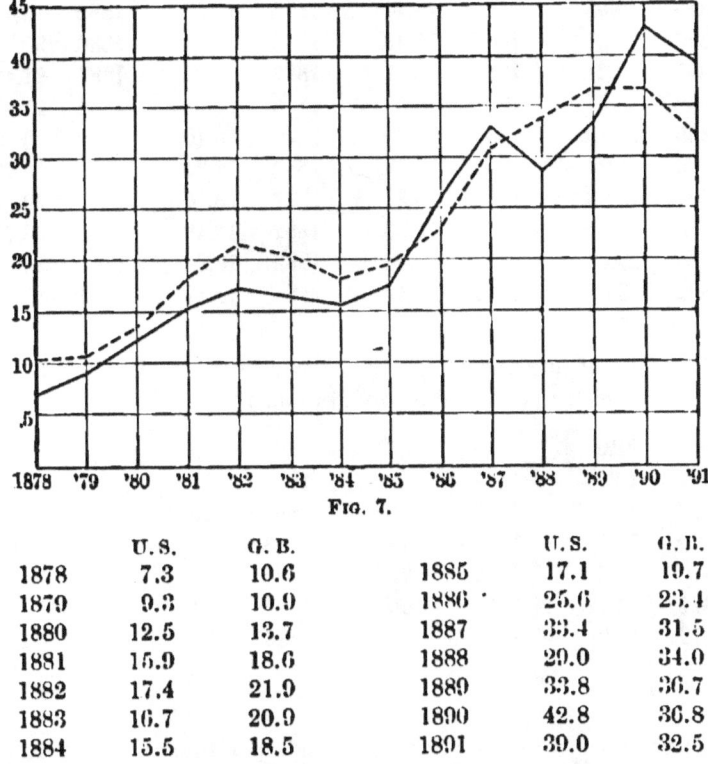

Fig. 7.

	U.S.	G.B.		U.S.	G.B.
1878	7.3	10.6	1885	17.1	19.7
1879	9.3	10.9	1886	25.6	23.4
1880	12.5	13.7	1887	33.4	31.5
1881	15.9	18.6	1888	29.0	34.0
1882	17.4	21.9	1889	33.8	36.7
1883	16.7	20.9	1890	42.8	36.8
1884	15.5	18.5	1891	39.0	32.5

The table furnishes a number of discrete points which in the figure are connected by straight lines to assist the eye.

Problems. — Exhibit graphically the information contained in the following tables:

1. Cost of steel rails per long ton in Pennsylvania mills from 1867 to 1894. (Mineral Resources.)

1867	$166.00	1874	$94.25	1881	$61.13	1888	$29.83
1868	158.50	1875	68.75	1882	48.50	1889	29.25
1869	132.25	1876	59.25	1883	37.75	1890	31.75
1870	106.75	1877	45.50	1884	30.75	1891	29.92
1871	102.50	1878	42.25	1885	28.50	1892	30.00
1872	112.00	1879	48.25	1886	34.50	1893	28.12
1873	120.50	1880	67.50	1887	37.08	1894	24.00

ANALYTIC GEOMETRY

2. Commercial value of one ounce gold in ounces silver from 1855 to 1894. (Report of Director of Mint.)

1855	15.38	1865	15.44	1875	16.59	1885	19.41
1856	15.38	1866	15.43	1876	17.88	1886	20.74
1857	15.27	1867	15.57	1877	17.22	1887	21.13
1858	15.38	1868	15.59	1878	17.94	1888	21.99
1859	15.19	1869	15.60	1879	18.40	1889	22.09
1860	15.29	1870	15.57	1880	18.05	1890	19.76
1861	15.50	1871	15.57	1881	18.16	1891	20.92
1862	15.35	1872	15.63	1882	18.19	1892	23.72
1863	15.37	1873	15.92	1883	18.64	1893	26.49
1864	15.37	1874	16.17	1884	18.57	1894	32.56

3. Expense of moving freight per ton mile on N.Y.C. & H.R.R.R. from 1866 to 1894. (Poor's Railway Manual.)

1866	¢2.16	1873	¢1.03	1880	¢.54	1887	¢.56
1867	1.95	1874	.98	1881	.56	1888	.59
1868	1.80	1875	.90	1882	.60	1889	.57
1869	1.40	1876	.71	1883	.68	1890	.54
1870	1.15	1877	.70	1884	.62	1891	.57
1871	1.01	1878	.60	1885	.54	1892	.54
1872	1.13	1879	.55	1886	.53	1893	.54
						1894	.57

4. Pressure of saturated steam in pounds per square inch at intervals of 9° from 32° to 428° Fahrenheit. (Based on Regnault's results.)

32°	.085 lbs.	131°	2.27 lbs.	230°	20.80 lbs.	329°	101.9 lbs.
41	.122	140	2.88	239	24.54	338	115.1
50	.173	149	3.62	248	28.83	347	129.8
59	.241	158	4.51	257	33.71	356	145.8
68	.333	167	5.58	266	39.25	365	163.3
77	.456	176	6.87	275	45.49	374	182.4
86	.607	185	8.38	284	52.52	383	203.3
95	.806	194	10.16	293	60.40	392	225.9
104	1.06	203	12.26	302	69.21	401	250.3
113	1.38	212	14.70	311	79.03	410	276.9
122	1.78	221	17.53	320	89.86	419	305.5
						428	336.3

RECTANGULAR COORDINATES

In these problems it is evident that theoretically there corresponds a determinate value of the ordinate to every value of the abscissa. Hence the ordinate is called a function of the abscissa, even though it may be impossible to express the relation between ordinate and abscissa by a formula or analytic function.

5. Suppose a body falling freely under gravity down a vertical guide wire to have a pencil attached in such a manner that the pencil traces a line on a vertical sheet of paper moving horizontally from right to left with a uniform velocity. To determine the relation between the distance the body falls and the time of falling.*

Take the vertical and horizontal lines through the starting point as axes of reference, and let 01, 12, 23, ..., be the equal distances through which the sheet of paper moves per second, the spaces 05, 510, ..., on the vertical axis represent 5 feet. Then the ordinate of any point of the line traced by the pencil represents the distance the body has fallen during the time represented by the abscissa of the point. Careful measurements show that the distance varies as the square of the time. Calling the distance s, the time t, the distance the body falls the first second $\frac{1}{2}g$, where g is found by experiment to be 32.16 feet, the relation between ordinate and abscissa of the line traced by the pencil is expressed by the proportion $\dfrac{s}{\frac{1}{2}g} = \dfrac{t^2}{1}$, which leads to the equation $s = \frac{1}{2}gt^2$. The curve and the equation express the same physical law, the one algebraically, the other geometrically.

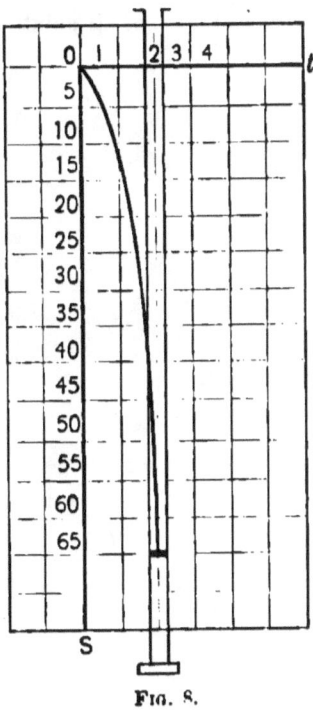

FIG. 8.

In this problem the ordinate is an analytic function of the abscissa, for the relation between the two is expressed by a formula.

The ordinate is a continuous function of the abscissa; that is, the difference between two ordinates can be made as small as we please by sufficiently diminishing the difference between the corresponding abscissas.

---

* This is the principle of Morin's apparatus for determining experimentally the law of falling bodies.

6. A body is thrown horizontally with a velocity of v feet per second. The only force disturbing the motion of the body taken into account is gravity. Find the position of the body t seconds after starting.

Calling the starting point the origin, the horizontal and vertical lines through the origin the X-axis and Y-axis respectively, the coordinates of the body t seconds after starting are $x = vt$, $y = -\frac{1}{2}gt^2$. Eliminating t, $y = -\frac{g}{2v^2}x^2$, an equation which expresses the relation existing between the coordinates of all points in the path of the body.

CHAPTER II

EQUATIONS OF GEOMETRIC FIGURES

Art. 7. — The Straight Line

A point moving in a plane generates either a straight line or a plane curve. Frequently the geometric law governing the motion of the point can be directly expressed in the form of an equation between the coordinates of the point. This equation is called the equation of the geometric figure generated by the point.

Draw a straight line through the origin. By elementary geometry $\frac{ba}{Aa} = \frac{b_1 a_1}{Aa_1} = \frac{b_2 a_2}{Aa_2} = \cdots$. This succession of equal ratios expresses a geometric property which characterizes points in the straight line; for every point in the line furnishes one of these ratios, and no point not in the straight line furnishes one of these ratios. Calling the common value of these ratios m, and letting x and y denote the coordinates of any point in the line, the equation $y = mx$ expresses the same geo-

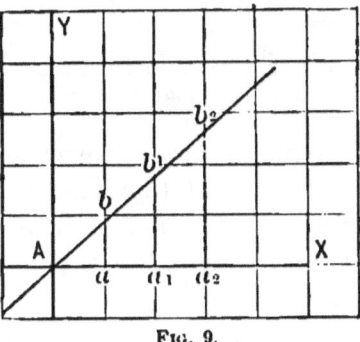

Fig. 9.

metric property as the succession of equal ratios. Hence if the point (x, y) is governed in its motion by the equation, it generates a straight line through the origin. By trigonometry m is the tangent of the angle through which the X-axis must be turned anti-clockwise to bring it into coincidence with the straight line.

This angle is called the angle which the line makes with the X-axis, and its tangent is called the slope of the line.

Give the straight line $y = mx$ a motion of translation parallel to the Y-axis upward through a distance n. The ordinate of every point in the line in the new position is n greater than the ordinate of the same point in the line through the origin.

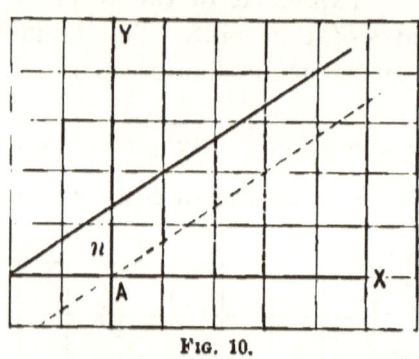

Fig. 10.

Hence the equation of the straight line, whose slope is m, and which intersects the Y-axis at a point n linear units above the origin, is $y = mx + n$. n is called the intercept of the line on the Y-axis. x and y are called the current coordinates of the straight line. m and n are called the parameters of the straight line. To every straight line there corresponds one pair of values of m and n; for a straight line makes only one angle with the X-axis, and intersects the Y-axis in only one point; conversely, to every pair of values of m and n there corresponds only one straight line.

Problems. — 1. Write the equation of the line parallel to the Y-axis at a distance of 5 linear units to the right of the Y-axis.

2. Write the equation of the line parallel to the X-axis intersecting the Y-axis 6 below the origin.

3. Write the equation of the straight line through the origin making an angle of 45° with the X-axis.

4. Find the equation of the line making an angle of 135° with the X-axis, intersecting the Y-axis 5 above the origin.

5. Write the equation of the line whose slope is 2, intercept on Y-axis -5.

6. Find the equation of the path of a point moving in such a manner that it is always equidistant from $(3, -5)$, $(-3, 5)$.

7. Find the equation of the path of a point moving in such a manner that it is always equidistant from $(4, 2)$, $(-3, 5)$.

8. Find the equation of the locus of the points equidistant from (7, 4), (− 3, − 5).

9. Find the equation of the straight line bisecting the line joining (2, − 5), (6, 3) at right angles.

Art. 8. — The Circle

According to the geometric definition of the circle the point (x, y) describes the circumference of a circle with radius r, center (a, b), if the point (x, y) moves in the XY-plane in such a manner that its distance from (a, b) is always r. This condition is expressed by the equation $(x-a)^2 + (y-b)^2 = r^2$, which is therefore the equation of a circle.

Problems. — **1.** Write the equation of the circle whose radius is 5, center (2, − 3).
2. Find the equation of the circle with center at origin, radius r.
3. Find equation of circle radius 5, center (5, 0).
4. Find equation of circle radius 5, center (5, 5).
5. Find equation of circle radius 5, center (− 5, 5).
6. Find equation of circle radius 5, center (− 5, − 5).
7. Find equation of circle radius 5, center (0, − 5).
8. Find equation of circle radius 5, center (0, 5).

Art. 9. — The Conic Sections

After studying the straight line and circle, the old Greek mathematicians turned their attention to a new class of curves which they called conic sections, because these curves were originally obtained by intersecting a cone by a plane. It was soon discovered that these curves may be defined thus:

A conic section is a curve traced by a point moving in a plane in such a manner that the ratio of the distances from the moving point to a fixed point and to a fixed line is constant.

This definition will be used to construct these curves, to obtain their properties, and to find their equations. The fixed point is called the focus, the fixed line the directrix of the conic section. When the constant ratio, called the characteristic ratio and denoted by e, is less than unity, the curve is called an ellipse; when greater than unity, an hyperbola; when equal to unity, a parabola.*

The following proposition is due to Quételet (1796–1874), a Belgian scholar:

If a right circular cone is cut by a plane, and two spheres are inscribed in the cone tangent to the plane, the two points of contact are the foci of the section of the cone by the plane; and the straight lines in which this plane is cut by the planes of the circles of contact of spheres and cone are the directrices corresponding to these two foci respectively.

Let the plane cut all the elements of one sheet of the cone. F, F' are the points of contact of the spheres with the cutting plane; P any point in the intersection of plane and surface of cone; T, T' the points of contact of element of cone through P with spheres. The plane of the elements Sa, Sa' is perpendicular to the cutting plane and the plane of the circles of contact. Since tangents from a point to a sphere are equal, $PF = PT$, $PF' = PT'$. Hence

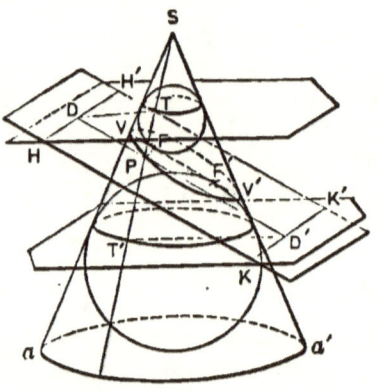

Fig. 11.

$$PF + PF' = PT + PT' = TT',$$

a constant. Through P draw DD' perpendicular to the parallels HH', KK'. From the similar triangles PDT and $PD'T'$, $\dfrac{PT}{PT'} = \dfrac{PD}{PD'}$, hence $\dfrac{PF}{PF'} = \dfrac{PD}{PD'}$. By composition $\dfrac{PF}{TT'} = \dfrac{PD}{DD'}$, by interchanging means, $\dfrac{PF}{PD} = \dfrac{TT'}{DD'}$, a constant. Similarly, $\dfrac{PF'}{TT'} = \dfrac{PD'}{DD'}$, $\dfrac{PF'}{PD'} = \dfrac{TT'}{DD'}$. Call the points

* Cayley, in the article on Analytic Geometry in the Britannica, ninth edition, calls this definition of conic sections the definition of Apollonius. Apollonius, a Greek mathematician, about 205 B.C., wrote a treatise on Conic Sections.

of intersection of the straight line FF'' with the section of the cone V, V''. Since

$$VF + VF'' = FF'' + 2\,VF = TT''$$

and
$$V''F + V''F'' = FF'' + 2\,V''F'' = TT'',$$
$$VF = V'F''$$

and
$$VF + VF'' = V''F + VF'' = VV' = TT''.$$

Hence the constant ratio $\dfrac{TT''}{DD'} = \dfrac{VV'}{DD'}$ is less than unity, and the conic section is an ellipse.

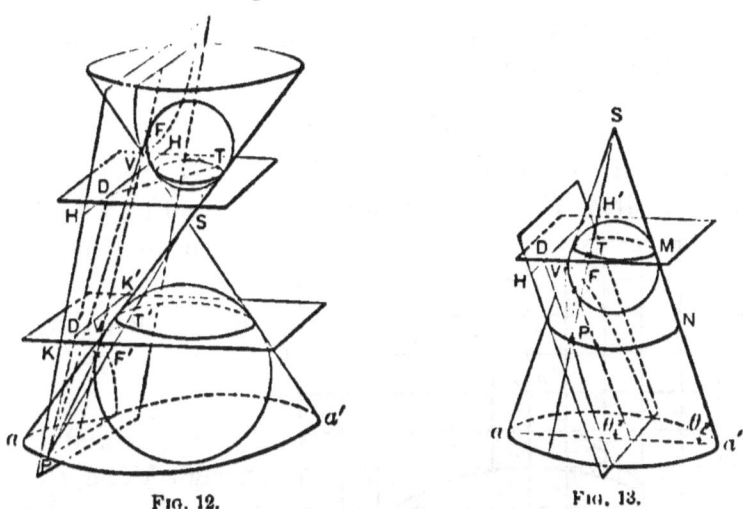

Fig. 12.   Fig. 13.

It is seen that the ellipse may also be defined as the locus of the points, the sum of whose distances from two fixed points, the foci, is constant.

Let the plane cut both sheets of the cone. With the same notation as before, $PF = PT$, $PF'' = PT''$; hence

$$PF - PF'' = TT'' = \text{a constant}.$$

From the similar triangles PDT and $PD'T''$, $\dfrac{PT}{PT''} = \dfrac{PD}{PD'}$; hence $\dfrac{PF}{PF''} = \dfrac{PD}{PD'}$. By division $\dfrac{PF}{TT''} = \dfrac{PD}{DD'}$; hence

$$\dfrac{PF}{PD} = \dfrac{TT''}{DD'} = \text{a constant}.$$

Similarly, $\dfrac{PF''}{PD'} = \dfrac{TT'}{DD'}$. $\dfrac{TT'}{DD'}$ is greater than unity, and the conic section is an hyperbola.

The hyperbola may also be defined as the locus of points, the difference of whose distances from the foci is constant.

Let the cutting plane and the element MN make the same angle θ with a plane perpendicular to the axis of the cone. The intersections of planes through the element MN with the cutting plane are perpendicular to the intersection of cutting plane with plane of circle of contact.

$PF' = PT = MN = PD$, and the conic section is a parabola, focus F', directrix HH'.

Art. 10. — The Ellipse, $e < 1$

Construction.—Let F be the focus, HH' the directrix. Through F draw FK perpendicular to HH', and on the perpendicular to FK through F take the points P and P'' such that

$$\dfrac{PF}{FK} = \dfrac{P''F}{FK} = e.$$

Through K and P, and through K and P'', draw straight lines. Draw any number of straight lines parallel to HH', intersecting KP and KP'' in $n_1, n_2, n_3, n_4, \ldots$, FK in $m_1, m_2, m_3, \ldots$. With F as center and $m_1 n_1$ as radius describe an arc intersecting $n_1 n_1$ in P_1. Then

$$\dfrac{P_1 F}{m_1 K} = \dfrac{m_1 n_1}{m_1 K} = \dfrac{PF}{FK} = e,$$

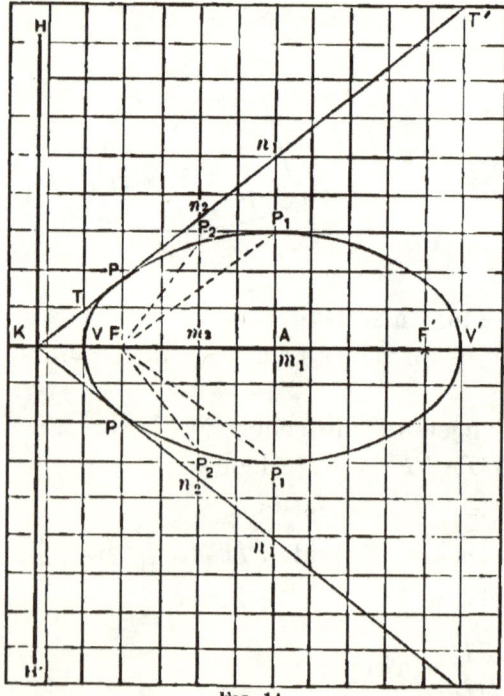

Fig. 14.

EQUATIONS OF GEOMETRIC FIGURES

and P_1 is a point in the ellipse. Similarly, an infinite number of points of the curve may be located.

Definitions. — The perpendicular through the focus to the directrix is called the axis of the ellipse. The axis intersects the curve in the points V and V', dividing FK internally and externally into segments whose ratio is e. The points V and V' are called the vertices, the point A midway between V and V', the center of the ellipse. The finite line VV' is the transverse axis or major diameter, denoted by $2a$; the line P_1P_1' perpendicular to VV' at A and limited by the curve is the conjugate axis or minor diameter, denoted by $2b$; the finite line PP' is called the parameter of the ellipse, denoted by $2p$. The lines KP and KP' are called focal tangents. The ratio of the distance from the focus to the center to the semi-major diameter is called the eccentricity of the ellipse.

Properties. — The foci F and F' are equidistant from the center A. By the definition of the ellipse $VF = e \cdot VK$, $V'F = e \cdot V'K$. Subtracting, $FF' = e \cdot VV'$. Dividing by 2, $AF = e \cdot AV$. Hence $e = \dfrac{AF}{a}$, that is, in the ellipse the eccentricity equals the characteristic ratio.

By definition $\dfrac{FP_1}{AK} = e$, and by construction

$$FP_1 = An_1 = \frac{VT + V'T'}{2} = \frac{VF + V'F}{2} = a. \quad \text{Hence } AK = \frac{a}{e}.$$

By definition eccentricity $e = \dfrac{AF}{AV} = \dfrac{\sqrt{a^2 - b^2}}{a}$.

From the figure, $VF = AV - AF = a - ae = a(1 - e)$;

$$V'F = AV' + AF = a + ae = a(1 + e).$$

VF and VF' are called the focal distances.

By definition $\dfrac{VF}{VK} = e$, hence $VK = \dfrac{a(1-e)}{e}$; $\dfrac{V'F'}{V'K} = e$, hence $V'K = \dfrac{a(1+e)}{e}$.

From the figure $FK = AK - AF = \dfrac{a}{e} - ae = \dfrac{a(1-e^2)}{e}$;

$$F'K = AK + AF' = \dfrac{a}{e} + ae = \dfrac{a(1+e^2)}{e}.$$

By definition

$\dfrac{p}{FK} = e$, hence $p = a(1-e^2) = a\left(1 - \dfrac{a^2-b^2}{a^2}\right) = a\dfrac{b^2}{a^2} = \dfrac{b^2}{a}$.

Equation. — Take the axis of the ellipse as X-axis, the perpendicular to the axis through the center as Y-axis. Let P be any point of the curve, its coordinates x and y. The problem is to express the definition $PF = e \cdot PH$ by means of an equation between x and y. The definition is equivalent to $\overline{PF}^2 = e^2 \cdot \overline{PH}^2$, which is the same as

$$PD^2 + (AD + AF)^2 = e^2(AK + AD)^2,$$

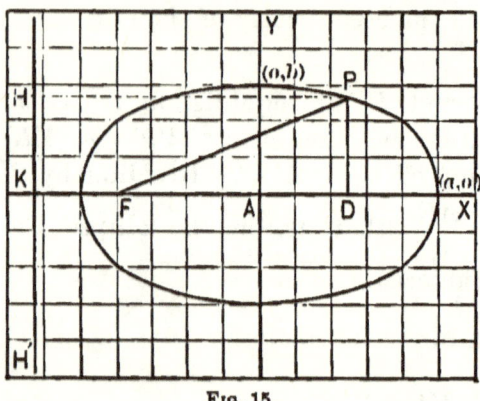

Fig. 15.

which becomes $\quad y^2 + (x + ae)^2 = e^2\left(\dfrac{a}{e} + x\right)^2$,

reducing to $\quad \dfrac{x^2}{a^2} + \dfrac{y^2}{a^2(1-e^2)} = 1$.

Since the point (o, b) is in the curve, $a^2(1-e^2) = b^2$, and the equation finally becomes $\dfrac{x^2}{a^2} + \dfrac{y^2}{b^2} = 1$.

Summary.— Collecting the results of the preceding paragraphs, the fundamental properties of the ellipse $\frac{x^2}{a^2} + \frac{y^2}{b^2} = 1$ are:

Distance from focus to extremity of conjugate diameter	a
Distance from center to directrix	$\dfrac{a}{e}$
Distance from focus to center	ae
Distance from focus to near vertex	$a(1-e)$
Distance from focus to far vertex	$a(1+e)$
Distance from directrix to near focus	$\dfrac{a(1-e^2)}{e}$
Distance from directrix to far focus	$\dfrac{a(1+e^2)}{e}$
Distance from directrix to near vertex	$\dfrac{a(1-e)}{e}$
Distance from directrix to far vertex	$\dfrac{a(1+e)}{e}$
Eccentricity	$e = \dfrac{(a^2-b^2)^{\frac{1}{2}}}{a}$
Square of semi-conjugate diameter	$b^2 = a^2(1-e^2)$
Semi-parameter	$p = a(1-e^2) = \dfrac{b^2}{a}$

Art. 11.— The Hyperbola, $e > 1$

Construction.— Draw FK through the focus F perpendicular to the directrix HH'. On the perpendicular to FK through F take the points P and P' such that $\dfrac{PF}{FK} = \dfrac{P'F}{FK} = e$. Through K and P, and through K and P' draw straight lines. Draw any number of parallels to HH', and on these parallels locate points of the curve exactly as was done in the ellipse. The hyperbola consists of two infinite branches. The vertices V and V' divide FK internally and externally into segments whose ratio is e. The construction shows that the parallels to HH' between V and V' do not contain points of the curve. The notation is the same as for the ellipse.

Properties.— From the definition of the hyperbola $VF = e \cdot VK$, $V''F = e \cdot V''K$. Adding $FF'' = e \cdot VV''$; dividing by 2, $AF = e \cdot AV$. Hence $e = \dfrac{AF}{a} =$ eccentricity, that is in the hyperbola also the characteristic ratio equals the eccentricity.

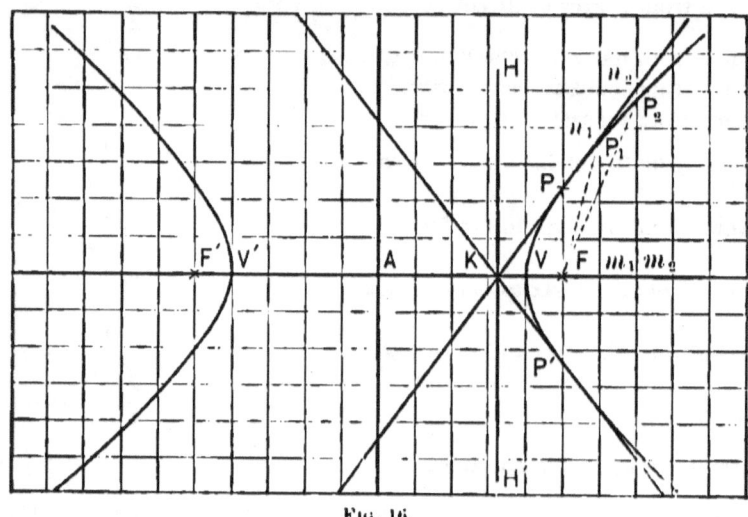

Fig. 16.

From the figure
$VF = AF - AV = a(e-1)$; $V'F = AF + AV' = a(e+1)$.

By definition
$\dfrac{VF}{VK} = e$, hence $VK = \dfrac{a(e-1)}{e}$; $\dfrac{V''F}{V''K} = e$, hence $V'K = \dfrac{a(e+1)}{e}$.

From the figure $AK = AV - VK = a - a\dfrac{e-1}{e} = \dfrac{a}{e}$.

From the figure
$FK = AF - AK = \dfrac{a(e^2-1)}{e}$; $F''K = AK + AF = \dfrac{a(e^2+1)}{e}$.

By definition $\dfrac{p}{FK} = e$, hence $p = a(e^2 - 1)$.

Equation.— To find the equation of the hyperbola take the axis of the curve as X-axis, the perpendicular to the axis through the center as Y-axis. Let P be any point of the curve, its coordinates x and y. The problem is to express the definition $PF = e \cdot PH$ by means of an equation between x and y.

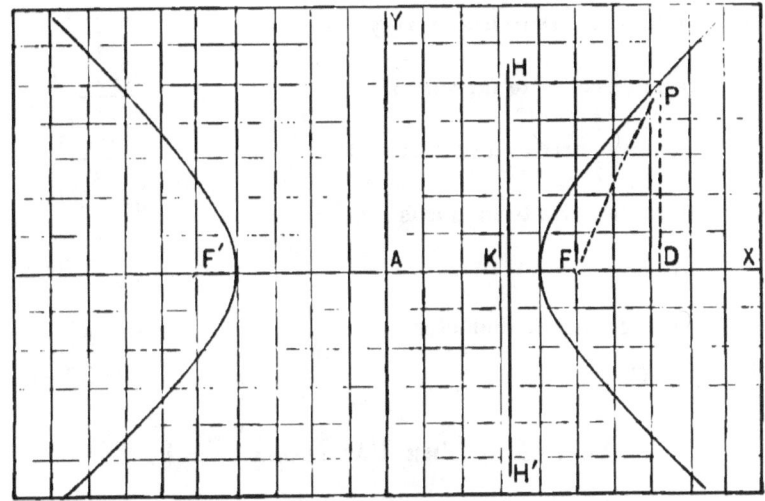

Fig. 17.

The definition is equivalent to $PF^2 = e^2 \cdot \overline{PH}^2$, which is the same as $PD^2 + (AD - AF)^2 = e^2(AD - AK)^2$, which becomes

$$y^2 + (x - ae)^2 = e^2\left(x - \frac{a}{e}\right)^2, \text{ reducing to } \frac{x^2}{a^2} - \frac{y^2}{a^2(e^2-1)} = 1.$$

Placing $a^2(e^2 - 1) = b^2$, the equation takes the form $\frac{x^2}{a^2} - \frac{y^2}{b^2} = 1$.

Since $b^2 = a^2e^2 - a^2$, it is seen from the figure that an arc described from the vertex as a center with a radius equal to distance from focus to center intersects the Y-axis at a distance b from the center. $2b$ is called the conjugate or minor diameter of the hyperbola.

Summary. — Collecting the results of the preceding paragraphs, the fundamental properties of the hyperbola

$$\frac{x^2}{a^2} - \frac{y^2}{b^2} = 1 \text{ are:}$$

ANALYTIC GEOMETRY

Distance from vertex to extremity of conjugate diameter	ae
Distance from focus to center	ae
Distance from focus to near vertex	$a(e-1)$
Distance from focus to far vertex	$a(e+1)$
Distance from directrix to near vertex	$\dfrac{a(e-1)}{e}$
Distance from directrix to far vertex	$\dfrac{a(e+1)}{e}$
Distance from directrix to near focus	$\dfrac{a(e^2-1)}{e}$
Distance from directrix to far focus	$\dfrac{a(e^2+1)}{e}$
Eccentricity	$e = \dfrac{(a^2+b^2)^{\frac{1}{2}}}{a}$
Square of semi-conjugate diameter	$b^2 = a^2(e^2-1)$
Semi-parameter	$p = a(e^2-1) = \dfrac{b^2}{a}$

Art. 12. — The Parabola, $e = 1$

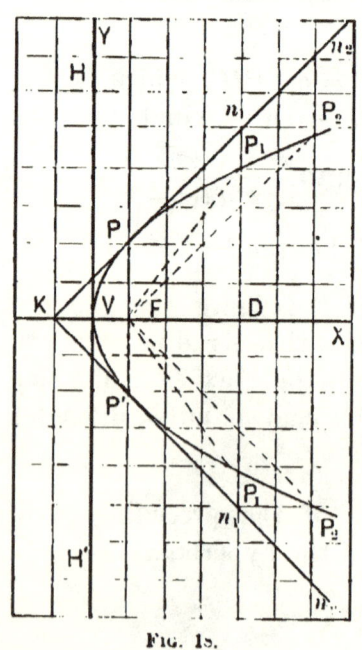

Fig. 18.

Through F, the focus, draw FK perpendicular to HH', the directrix. On the perpendicular to FK through F lay off $FP = FP'' = FK$. Draw the focal tangents KP and KP'', draw a series of parallels to HH', and locate points of the parabola as in the case of ellipse and hyperbola. From the figure, it is seen that the distance from vertex to focus is $\frac{1}{2}p$, distance from vertex to directrix is $\frac{1}{2}p$, the parameter being $2p$.

To find the equation of the parabola, take the axis of the curve as X-axis, the perpendicular to the axis at the vertex as Y-axis. Let $P(x, y)$ be any point in the curve.

The problem is to express the definition $PF = DK$ by means of an equation between x and y. The definition may be written $PF^2 = DK^2$, which is the same as $PD^2 + FD^2 = (VD + VK)^2$, which becomes

$$y^2 + (x - \tfrac{1}{2}p)^2 = (x + \tfrac{1}{2}p)^2,$$

reducing to $y^2 = 2px$.

A parabola whose focus and directrix are known may be generated mechanically as indicated in the figure.

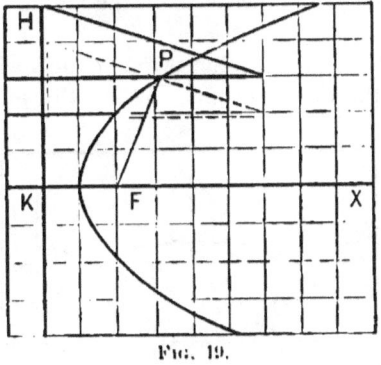

Fig. 19.

Problems. — 1. Construct the ellipse whose parameter is 6, eccentricity $\tfrac{2}{3}$.

2. Construct the hyperbola whose parameter is 8, eccentricity $\tfrac{3}{2}$.

3. Construct the ellipse whose diameters are 10 and 8. Find the equation of the ellipse, its eccentricity, and parameter.

4. Construct the hyperbola whose diameters are 8 and 6. Find the equation of the hyperbola, its eccentricity, and parameter.

5. Construct the parabola whose parameter is 12 and find its equation.

6. Find the equation of the ellipse whose eccentricity is $\tfrac{2}{3}$, major diameter 10.

7. The diameters of an hyperbola are 10 and 6. Find distances from center to focus and directrix.

8. The distances from focus to vertices of an hyperbola are 10 and 2. Find diameters.

9. The parameter of a parabola is 12. Find distance from focus to point in curve whose abscissa is 8.

10. Find diameters of the ellipse whose parameter is 16, eccentricity $\tfrac{3}{4}$.

11. In an ellipse, the distance from vertex to directrix is 6, eccentricity $\tfrac{1}{2}$. Find diameters and construct ellipse.

12. In the ellipse $\dfrac{x^2}{a^2} + \dfrac{y^2}{b^2} = 1$ show that the distances from the foci to the point (x, y) are $r = a - ex$, $r' = a + ex$. r and r' are called the focal radii of the point (x, y). The sum of the focal radii of the ellipse is constant and equal to $2a$.

13. In the hyperbola $\dfrac{x^2}{a^2} - \dfrac{y^2}{b^2} = 1$ show that the focal radii of the point (x, y) are $r = ex - a$, $r' = ex + a$. The constant difference of the focal radii of the hyperbola is $2a$.

14. Find the equation of the ellipse directly from the definition: The ellipse is the locus of the points the sum of whose distances from the foci equals $2a$.

Take the line through the foci as X-axis, the point midway between the foci as origin. When the point (x, y) is on the Y-axis its distances from F and F' are each equal to a. Call $AF = AF' = c$, the distance of (x, y) when on the Y-axis from the origin b. Then $a^2 - c^2 = b^2$. The geometric condition $PF + PF' = 2a$ is expressed by the equation

$$\sqrt{y^2 + (x-c)^2} + \sqrt{y^2 + (x+c)^2} = 2a,$$

which reduces to $\dfrac{x^2}{a^2} + \dfrac{y^2}{b^2} = 1$.

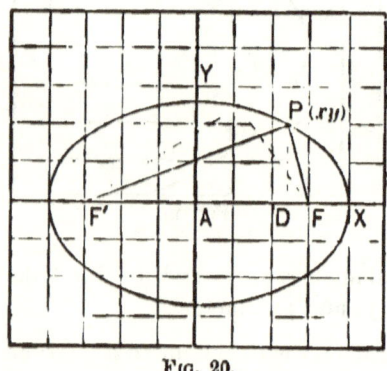

Fig. 20.

The definition used in this problem suggests a very simple mechanical construction of the ellipse whose foci and major diameter are known. Fasten the ends of an inextensible string of constant length $2a$ at the foci F and F'. A pencil point guided in the plane by keeping the string stretched traces the ellipse.

15. Find the equation of the hyperbola directly from the definition: The hyperbola is the locus of the points the difference of whose distances from the foci is $2a$.

Take the line through the foci as X-axis, the point midway between the foci as origin. Call $AF = AF' = c$, $c^2 - a^2 = b^2$.

The condition $PF' - PF = 2a$ leads to the equation

$$\sqrt{y^2 + (x+c)^2} - \sqrt{y^2 + (x-c)^2} = 2a,$$

which reduces to $\dfrac{x^2}{a^2} - \dfrac{y^2}{b^2} = 1$.

The mechanical construction of the hyperbola is effected as indicated in the figure.

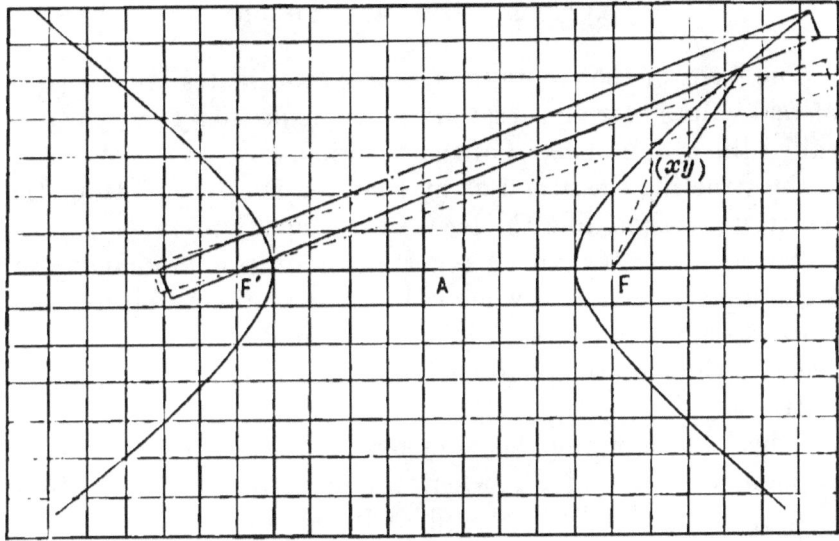

Fig. 21.

16. Two pins fixed in a ruler are constrained to move in grooves at right angles to each other. Show that every point of the ruler describes an ellipse whose semi-diameters are the distances from the point to the pins. This device is called an elliptic compass.

The following statements may help to form an idea of the importance of the conic sections:

The planets and asteroids move in ellipses with the sun at one focus.

The eccentricity of the earth's orbit is about $\frac{1}{60}$.

The eccentricity of the moon's elliptic path about the earth is about $\frac{1}{18}$.

Nearly all comets move in parabolas with the sun at the focus.

The cable of a suspension bridge, if the load is uniformly distributed over the horizontal, takes the form of a parabola.

A projectile, unless projected vertically, moves in a parabola, if the earth's attraction is the only disturbing force taken into account.

CHAPTER III

PLOTTING OF ALGEBRAIC EQUATIONS

Art. 13.—General Theory

The locus of the points (x, y) whose coordinates are the pairs of real values of x and y satisfying the equation $f(x, y) = 0$ is called the graph or locus of the equation.

Constructing the graph of an equation is called plotting the equation, or sketching the locus of the equation.

An equation $f(x, y) = 0$ is an algebraic equation, and y an algebraic function of x, when only the operations addition, subtraction, multiplication, division, involution, and evolution occur in the equation, and each of these only a finite number of times.

When the equation has the form $y = f(x)$, y is called an explicit function of x; when the equation has the form $f(x, y) = 0$, y is called an implicit function of x.

The locus represented by an equation $f(x, y) = 0$ depends on the relative values of the coefficients of the equation. For $mf(x, y) = 0$, where m is any constant, is satisfied by all the pairs of values of x and y which satisfy $f(x, y) = 0$, and by no others.

If the graphs of two equations $f_1(x, y) = 0$, $f_2(x, y) = 0$ are constructed, the coordinates of the points of intersection of these graphs are the pairs of real values of x and y which satisfy $f_1(x, y) = 0$ and $f_2(x, y) = 0$ simultaneously.

Occasionally it is possible to obtain the geometric definition of a locus directly from its equation, and then construct the locus mechanically. The equation $x^2 + y^2 = 25$ is at once seen

PLOTTING OF ALGEBRAIC EQUATIONS

to represent a circle with center at origin, radius 5. In general it is necessary to locate point after point of the locus by assigning arbitrary values to one variable, and computing the corresponding values of the other from the equation.

Art. 14. — Locus of First Degree Equation

The locus of the general first degree equation between two variables x and y, $Ax + By + C = 0$, is the locus of

$$y = -\frac{A}{B}x - \frac{C}{B}.$$

Moving the locus represented by this equation parallel to the Y-axis upward through a distance $\frac{C}{B}$, increases each ordinate by $\frac{C}{B}$. Hence the equation of the locus in the new position is $y = -\frac{A}{B}x$, which represents a straight line through the origin, since the ordinate is proportional to the abscissa. The equation $Ax + By + C = 0$ therefore represents a straight line whose slope is $-\frac{A}{B}$, and whose intercept on the Y-axis is $-\frac{C}{B}$. The intercept of this line on the X-axis, found by placing y equal to zero in the equation and solving for x, is $-\frac{C}{A}$.

The straight line represented by a first degree equation may be constructed by determining the point of intersection of the line with the Y-axis and the slope of the line, by determining any point of the line and the slope of the line, by determining the points of intersection of the line with the coordinate axes, by locating any two points of the line.

Problems. — Construct by the different methods the lines represented by the equations.

1. $2x + 3y = 6$.
2. $y = x - 5$.
3. $\frac{1}{2}x - \frac{1}{3}y = 1$.
4. $\frac{1}{3}x - \frac{1}{4}y = 2$.
5. $\frac{x}{2} + \frac{y}{3} = 1$.
6. $\frac{x}{a} + \frac{y}{b} = 1$.

7. Show that $y^2 - 2xy - 8x^2 = 0$ represents two straight lines through the origin.

8. Show that a homogeneous equation of the nth degree between x and y represents n straight lines through the origin.

9. Construct the straight line $\dfrac{x}{2} + \dfrac{y}{3} = 1$ and the circle $x^2 + y^2 = 25$ and compute the coordinates of the points of intersection. Verify by measurement.

Art. 15. — Straight Line through a Point

Through the fixed point (x_0, y_0) draw a straight line making an angle α with the X-axis. Let (x, y) be any point of this line, d the distance of (x, y) from (x_0, y_0). From the figure $x - x_0 = d \cos \alpha$, $y - y_0 = d \sin \alpha$, whence $x = x_0 + d \cos \alpha$,

$$y = y_0 + d \sin \alpha,$$

and $y - y_0 = \tan \alpha (x - x_0)$. That is, if a point (x, y) is governed in its motion by the equation

$$y - y_0 = \tan \alpha (x - x_0),$$

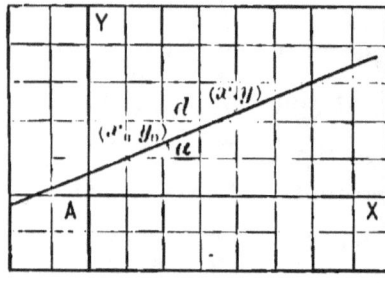

Fig. 22.

it generates a straight line through (x_0, y_0), making an angle α with the X-axis, and the coordinates of the point in this line at a distance d from (x_0, y_0) are $x = x_0 + d \cos \alpha$, $y = y_0 + d \sin \alpha$. The distance, d, is positive when measured from (x_0, y_0) in the direction of the side of the angle α through (x_0, y_0); negative when measured in the opposite direction.

Problems. — 1. Express the coordinates of a point in the straight line through $(2, 3)$ making an angle of $30°$ with the X-axis in terms of the distance from $(2, 3)$ to the point.

2. On the straight line through $(3, -2)$ making an angle of $60°$ with the X-axis, find the coordinates of the points whose distances from $(3, -2)$ are 10 and -10.

3. Write the equation of the straight line through $(-2, 5)$ and making an angle of $45°$ with the X-axis.

4. Write the equation of the straight line through $(4, -1)$ whose slope is $\frac{1}{2}$.

5. Find the distances from the point $(2, 3)$ to the points of intersection of the line through this point, making an angle of $30°$ with the X-axis and the circle $x^2 + y^2 = 25$.

The coordinates of any point of the given line are $x = 2 + d\cos 30°$, $y = 3 + d\sin 30°$. These values of x and y substituted in the equation of the circle $x^2 + y^2 = 25$ give the equation $d^2 + (4\cos 30° + 6\sin 30°)d = 12$, which determines the values of d for the points of intersection.

Art. 16. — Tangents

To plot a numerical algebraic equation involving two variables, put it into the form $y = f(x)$ if possible. Compute the values of y for different values of x, and locate the points whose coordinates are the pairs of corresponding real values of x and y. Connect the successive points by straight lines, and observe the form towards which the broken line tends, as the number of points located is indefinitely increased. This limit of the broken line is the locus of the equation.

EXAMPLE. — Plot $y^2 + x^2 = 9$.

Here $y = \pm\sqrt{9 - x^2}$, $x = \pm\sqrt{9 - y^2}$.

$x =$	-4	-3	-2	-1
$y = $	$\pm\sqrt{-7}$	0	$\pm\sqrt{5}$	$\pm 2\sqrt{2}$
	0	$+1$	$+2$	$+3$	$+4$
	± 3	$\pm 2\sqrt{2}$	$\pm\sqrt{5}$	0	$\pm\sqrt{-7}$,

or extracting the roots

$x =$	-3	-2	-1	0	$+1$	$+2$	$+3$
$y =$	0	± 2.237	± 2.828	± 3	± 2.828	± 2.237	0

y has two numerically equal values for each value of x. Hence the locus is symmetrical with respect to the X-axis. For a like reason the locus is symmetrical with respect to the

Y-axis. For values of $x > +3$ and for values of $x < -3$, y is imaginary. Hence the curve lies between the lines $x = +3$, $x = -3$. The curve also lies between the lines $y = +3$,

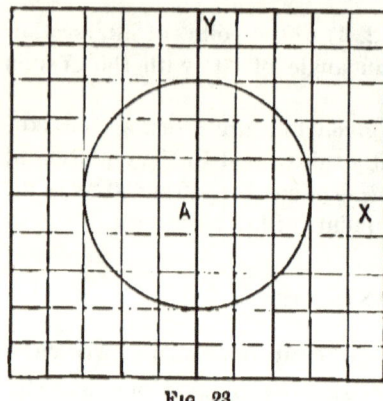

Fig. 23.

$y = -3$. Locating points of the locus and connecting them by straight lines, the figure formed approaches a circle more and more closely as the number of points located is increased. The form of the equation shows at once that the locus is a circle whose radius is 3, center the origin.

Through a point (x_0, y_0) of the circle an infinite number of straight lines may be drawn. The coordinates of any point of the straight line $y - y_0 = \tan \alpha (x - x_0)$ through (x_0, y_0), making an angle α with the X-axis are $x = x_0 + d \cos \alpha$, $y = y_0 + d \sin \alpha$. The point (x, y) is a point of the circle $x^2 + y^2 = 9$ when

$$(x_0 + d \cos \alpha)^2 + (y_0 + d \sin \alpha)^2 = 9,$$

that is, when

(1) $(x_0^2 + y_0^2 - 9) + 2 (\cos \alpha \cdot x_0 + \sin \alpha \cdot y_0) d + d^2 = 0.$

Equation (1) determines two values of d, and to each of these values of d there corresponds one point of intersection of line

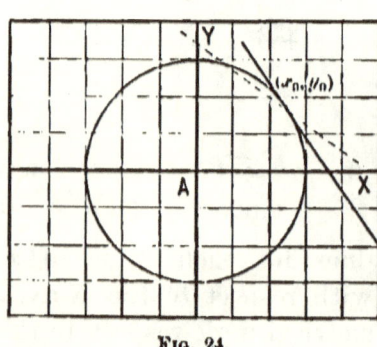

Fig. 24.

and circle. Since the point (x_0, y_0) is in the circle $x^2 + y^2 = 9$, the first term of equation (1) is zero, hence the equation has two roots equal to zero when

$$\cos \alpha \cdot x_0 + \sin \alpha \cdot y_0 = 0,$$

that is, when $\tan \alpha = -\dfrac{x_0}{y_0}$. To $d = 0$ there corresponds the point (x_0, y_0), and when both roots of equation (1) are zero, the two points of intersection of the straight line $y - y_0 = \tan\alpha\,(x - x_0)$ and the circle $x^2 + y^2 = 9$ coincide at (x_0, y_0), and the line is the tangent to the circle (x_0, y_0). Hence the equation of the tangent to the circle $x^2 + y^2 = 9$ at the point (x_0, y_0) is

$$y - y_0 = -\frac{x_0}{y_0}(x - x_0),$$

which reduces to $xx_0 + yy_0 = 9$.

A tangent to any curve is defined as a secant having two points of intersection with the curve coincident.* By the direction of the curve at any point is meant the direction of the tangent to the curve at the point.

The circle $x^2 + y^2 = 9$ at the point (x_0, y_0) makes, with the X-axis, $\tan^{-1}\left(-\frac{x_0}{y_0}\right)$. At the points corresponding to $x = 2$ the angles are $\tan^{-1}\left(-\frac{2}{\sqrt{5}}\right) = 138°\,37'$ and $\tan^{-1}\left(-\frac{2}{\sqrt{5}}\right) = 41°\,23'$.

Problems. — **1.** Show that $\dfrac{xx_0}{a^2} + \dfrac{yy_0}{b^2} = 1$ is tangent to the ellipse

$$\frac{x^2}{a^2} + \frac{y^2}{b^2} = 1 \text{ at } (x_0, y_0).$$

2. Show that $\dfrac{xx_0}{a^2} - \dfrac{yy_0}{b^2} = 1$ is tangent to the hyperbola

$$\frac{x^2}{a^2} - \frac{y^2}{b^2} = 1 \text{ at } (x_0, y_0).$$

3. Show that $yy_0 = p(x + x_0)$ is tangent to the parabola $y^2 = 2px$ at (x_0, y_0).

Art. 17. — Points of Discontinuity

EXAMPLE. — Plot $y = \dfrac{x+1}{x-2}$.

$x = -\infty \;\cdots\; -5 \;\; -4 \;\; -3 \;\; -2 \;\; -1 \;\; 0 \;\; +1 \;\; +2 \;\; +3 \;\; +4 \;\cdots\; +\infty$

$y = +1 \;\cdots\; +\tfrac{4}{7} \;\; +\tfrac{3}{6} \;\; +\tfrac{2}{5} \;\; +\tfrac{1}{4} \;\; 0 \;\; -\tfrac{1}{2} \;\; -2 \;\; \mp\infty \;\; +4 \;\; +2\tfrac{1}{2} \;\cdots\; +1$

* The secant definition of a tangent is due to Descartes and Fermat.

From $x = 0$ to $x = +2$, y is negative and increases indefinitely in numerical value as x approaches 2. From $x = +2$ to $x = +\infty$, y is positive and diminishes from $+\infty$ to $+1$.

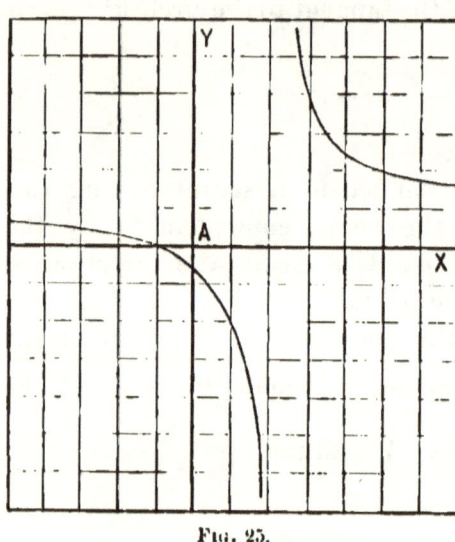

Fig. 25.

y is negative, and decreases numerically from $-\frac{1}{2}$ to 0 while x passes from 0 to -1. y is positive and increases to $+1$ from $x = -1$ to $x = -\infty$. The curve meets each of the two straight lines $x = 2$ and $y = 1$ at two points infinitely distant from the origin.

The point corresponding to $x = 2$ is a point of discontinuity of the curve. For if two abscissas are taken, one less than 2, the other greater than two, the difference between the corresponding ordinates approaches infinity when the difference between the abscissas is indefinitely diminished, while the definition of continuity requires that the difference between two ordinates may be made less than any assignable quantity by sufficiently diminishing the difference between the corresponding abscissas.

Art. 18. — Asymptotes

EXAMPLE. — Plot $y^2 - x^2 = 4$. Here $y = \pm \sqrt{x^2 + 4}$.

$x = -\infty \cdots -4 \quad -3 \quad -2 \quad -1 \quad 0 \quad +1 \quad +2$
$\quad +3 \quad +4 \cdots +\infty$

$y = \pm \infty \cdots \pm 4.47 \quad \pm 3.61 \quad \pm 2.83 \quad \pm 2.24 \quad \pm 2 \quad \pm 2.24 \quad \pm 2.83$
$\quad \pm 3.61 \quad \pm 4.47 \cdots \pm \infty$

PLOTTING OF ALGEBRAIC EQUATIONS

y has two numerically equal real values with opposite signs for every value of x. The values of y increase indefinitely in numerical value as x increases indefinitely in numerical value. It now becomes important to determine whether, as was the case in Art. 17, a straight line can be drawn which meets the curve in two points infinitely distant from the origin. The points of intersection of the straight line $y = mx + n$ and the locus of $y^2 - x^2 = 4$ are found by making

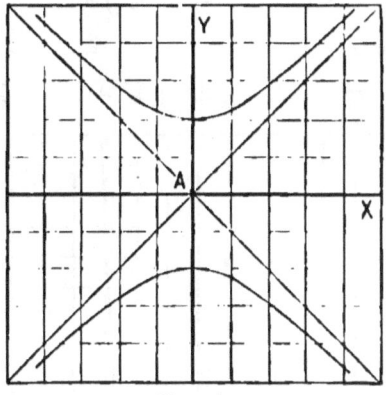

Fig. 26.

these equations simultaneous. Eliminating y, there results the equation in x, $(m^2 - 1)x^2 + 2mnx + n^2 - 4 = 0$. The problem is so to determine m and n that this equation has two infinite roots. An equation has two infinite roots when the coefficients of the two highest powers of the unknown quantity are zero.* Hence $y = mx + n$ meets $y^2 - x^2 = 4$ at two points infinitely distant from the origin when $m^2 - 1 = 0$, $2mn = 0$, whence $m = \pm 1$, $n = 0$. There are, therefore, two straight lines $y = x$ and $y = -x$, each of which meets the locus of $y^2 - x^2 = 4$ at two points infinitely distant from the origin. These lines are called asymptotes to the curve.

Problem. — Show that $y = \pm \dfrac{b}{a} x$ are asymptotes to the hyperbola

$$\frac{x^2}{a^2} - \frac{y^2}{b^2} = 1.$$

* Place $x = \dfrac{1}{z}$ in (1) $ax^n + bx^{n-1} + cx^{n-2} + \cdots + kx^2 + lx + m = 0$. There results (2) $a + bz + cz^2 + \cdots + kz^{n-2} + lz^{n-1} + mz^n = 0$. Equation (2) has two zero roots when $a = 0$, $b = 0$. Hence equation (1) has two infinite roots when $a = 0$, $b = 0$.

Art. 19. — Maximum and Minimum Ordinates

EXAMPLE. — Plot $y = x^3 - 7x + 7$.

$x = -\infty \cdots -4 \quad -3 \quad -2 \quad -1 \quad 0 \quad +1 \quad +1\tfrac{1}{2} \quad +2 \quad +3 \cdots +\infty$

$y = -\infty \cdots -29 \quad +1 \quad +13 \quad +13 \quad +7 \quad +1 \quad -\tfrac{1}{8} \quad +1 \quad +13 \cdots +\infty$

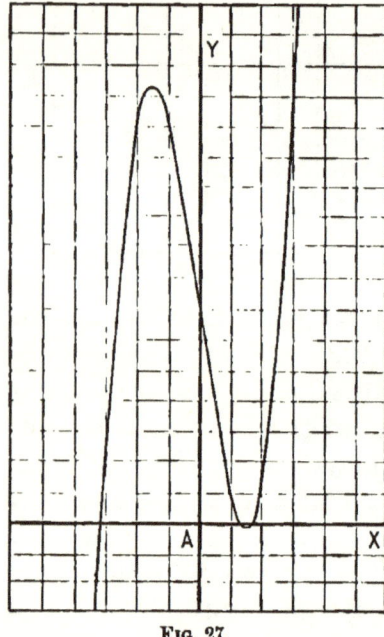

Fig. 27.

For $x = +1$, $y = +1$; for $x = +2$, $y = +1$; for $x = 1\tfrac{1}{2}$, $y = -\tfrac{1}{8}$. Hence between $x = 1$ and $x = 2$ the curve passes below the X-axis, turns and again passes above the X-axis. At the turning point the ordinate has a minimum value; that is, a value less than the ordinates of the points of the curve just before reaching and just after passing the turning point. The point generating the curve moves upward from $x = 0$ to $x = -1$, but somewhere between $x = -1$ and $x = -2$ the point turns and starts moving towards the X-axis. At this turning point the ordinate is a maximum; that is, greater than the ordinates of the points next the turning point on either side.

To determine the exact position of the turning points, let x' be the abscissa, y' the ordinate of the turning point. Let h be a very small quantity, y_1 the value of y corresponding to $x = x' \pm h$. Then $y_1 - y'$ must be positive when y' is a minimum, negative when y' is a maximum. Now

$$y_1 - y' = (3x^2 - 7)(\pm h) + 3x(\pm h)^2 + (\pm h)^3.$$

h may be taken so small that the lowest power of h determines the sign of $y_1 - y'$.* $y_1 - y'$ can therefore have the same sign for $\pm h$ only when the coefficient of the first power of h vanishes. This gives $3x^2 - 7 = 0$, whence $x = \pm \sqrt{\tfrac{7}{3}}$. $x = +\sqrt{\tfrac{7}{3}}$, rendering $y_1 - y'$ positive for $\pm h$, corresponds to a minimum ordinate; $x = -\sqrt{\tfrac{7}{3}}$, rendering $y_1 - y'$ negative for $\pm h$, corresponds to a maximum ordinate.

When $x = +\sqrt{\tfrac{7}{3}}$, $y = 7 - \tfrac{14}{9}\sqrt{21} = -.2$; when $x = -\sqrt{\tfrac{7}{3}}$, $y = 7 + \tfrac{14}{9}\sqrt{21} = 14.2$.†

The values of x which make $y = 0$ are the roots of the equation $x^3 - 7x + 7 = 0$. These values of x are the abscissas of the points where the locus of $y = x^3 - 7x + 7$ intersects the X-axis. $x^3 - 7x + 7 = 0$, therefore, has two roots between $+1$ and $+2$, and a negative root between -3 and -4.

Art. 20. — Points of Inflection

Example. — Plot $y + x^2y - x = 0$. Here $y = \dfrac{x}{1 + x^2}$.

$x = -\infty \;\cdots\; -3 \;\; -2 \;\; -1 \;\; 0 \;\; +1 \;\; +2 \;\; +3 \;\cdots\; +\infty$

$y = \;\;\; 0 \;\cdots\; -\tfrac{3}{10} \;\; -\tfrac{2}{5} \;\; -\tfrac{1}{2} \;\; 0 \;\; +\tfrac{1}{2} \;\; +\tfrac{2}{5} \;\; +\tfrac{3}{10} \;\cdots\; +0$

If (x, y) is a point of the locus, $(-x, -y)$ is also a point of the locus. Hence the origin is a center of symmetry of the locus. A line may be drawn through the origin intersecting the curve in the symmetrical points P and P'. If this line is

---

* Let $s = ah^3 + bh^4 + ch^5 + dh^6 + \cdots$ be an infinite series with finite coefficients, and let B be greater numerically than the largest of the coefficients $b, c, d, \cdots$. Then $bh + ch^2 + dh^3 + \cdots < h\,\dfrac{B}{1-h}$ and

$s = h^3(a + bh + ch^2 + dh^3 \cdots) = h^3(a \pm A)$, when $A < h\,\dfrac{B}{1-h}$.

When h is indefinitely diminished, $h\,\dfrac{B}{1-h}$ diminishes indefinitely. Consequently A becomes less than the finite quantity a, and s has the sign of ah^3.

† This method of examining for maxima and minima was invented by Fermat (1590-1663).

turned about A until P coincides with A, P' must also coincide with A. The line through A now becomes a tangent to the curve, but this tangent intersects the curve. From the figure it is seen that the coincidence of three points of intersection

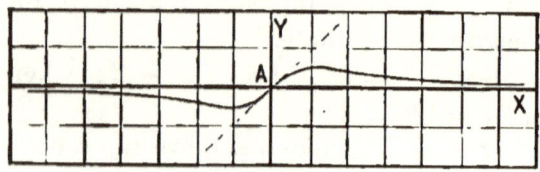

Fig. 28.

at the point of tangency, and the consequent intersection of the curve by the tangent, is caused by the fact that at the origin the curve changes from concave up to convex up. Such a point of the curve is called a point of inflection.

To find the analytic condition which determines a point of inflection, let (x_0, y_0) be any point of $y + x^2y - x = 0$. The coordinates of any point on a line through (x_0, y_0) are $x = x_0 + d\cos\alpha$, $y = y_0 + d\sin\alpha$. The points of intersection of line and curve correspond to the values of d satisfying the equation

$$(y_0 + x_0^2 y_0 - x_0) + (\sin\alpha - \cos\alpha + 2\cos\alpha \cdot x_0 y_0 + \sin\alpha \cdot x_0^2)d$$
$$+ (\cos^2\alpha \cdot y_0^2 + 2\sin\alpha\cos\alpha \cdot x_0)d^2 + \cos^2\alpha \sin\alpha \cdot d^3 = 0.$$

The first term of this equation vanishes by hypothesis, and if the coefficients of d and d^2 also vanish, the straight line and curve have three coincident points of intersection at (x_0, y_0). The simultaneous vanishing of the coefficients of d and d^2 requires that the equations

$$\sin\alpha - \cos\alpha + 2\cos\alpha \cdot x_0 y_0 + \sin\alpha \cdot x_0^2 = 0$$
and
$$\cos^2\alpha \cdot y_0^2 + 2\sin\alpha\cos\alpha \cdot x_0 = 0$$

determine the same value for $\tan\alpha$. This gives the equation $\dfrac{1 - 2x_0 y_0}{1 + x_0^2} = -\dfrac{y_0}{2x_0}$, reducing to $y_0 - 3x_0^2 y_0 + 2x_0 = 0$, which together with $y_0^2 + x_0^2 y_0 - x_0 = 0$ determines the three points of inflection $(0, 0)$, $(\sqrt{3}, \frac{1}{4}\sqrt{3})$, $(-\sqrt{3}, -\frac{1}{4}\sqrt{3})$.

Art. 21. — Diametric Method of Plotting Equations

Example. — Plot

$$y^2 - 2xy + 3x^2 - 16x = 0.$$

Here $y = x \pm \sqrt{16x - 2x^2}$.

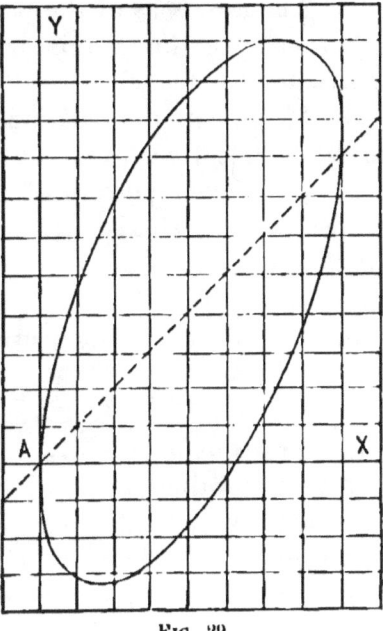

Fig. 29.

Draw the straight line $y = x$. Adding to and subtracting from the ordinate of this line, corresponding to any abscissa x, the quantity $\sqrt{16x - 2x^2}$, the corresponding ordinates of the required locus are obtained.

This locus intersects the line $y = x$ when $\sqrt{16x - 2x^2} = 0$, that is when $x = 0$ and $x = 8$. y is real only for values of x from 0 to 8. The curve intersects the X-axis when $x = \sqrt{16x - 2x^2}$, that is when $x = 5\frac{1}{3}$. Points of the curve are located by the table,

$x =$	0	+1	+2	+3
$\sqrt{16x - 2x^2} =$	0	$\pm\sqrt{14}$	$\pm 2\sqrt{6}$	$\pm\sqrt{30}$
	+4	+5	+6	+7	+8
	$\pm 4\sqrt{2}$	$\pm\sqrt{30}$	$\pm 2\sqrt{6}$	$\pm\sqrt{14}$	0

Art. 22. — Summary of Properties of Loci

From the discussions in the preceding articles, the following conclusions are obtained:

1. If the absolute term of an equation is zero, the origin is a point of the locus of the equation.

2. To find where the locus of an equation intersects the X-axis, place $y = 0$ in the equation and solve for x; to find where the locus intersects the Y-axis, place $x = 0$ and solve for y.

3. The abscissas of the points of intersection of the locus of $y = f(x)$ with the X-axis are the real roots of the equation $f(x) = 0$.

4. If the equation contains only even powers of y, the locus is symmetrical with respect to the X-axis; if the equation contains only even powers of x, the locus is symmetrical with respect to the Y-axis. The origin is a center of symmetry of the locus when $(-x, -y)$ satisfies the equation, because (x, y) does.

5. The points of intersection of the straight line

$$y - y_0 = \tan \alpha (x - x_0)$$

with the locus of $f(x, y) = 0$ are the points (x, y) corresponding to the values of d which are the roots of the equation obtained by substituting $x = x_0 + d \cos \alpha$, $y = y_0 + d \sin \alpha$ in $f(x, y) = 0$. The number of points of intersection is equal to the degree of the equation, and is called the order of the curve.

6. The distances from any point (x_0, y_0) to the points of intersection of the straight line $y - y_0 = \tan \alpha (x - x_0)$ with the locus of $f(x, y) = 0$ are the values of d which are the roots of the equation obtained by substituting $x = x_0 + d \cos \alpha$, $y = y_0 + d \sin \alpha$ in $f(x, y) = 0$.

7. The tangent to $f(x, y) = 0$ at (x_0, y_0) in the locus is found by substituting $x = x_0 + d \cos \alpha$, $y = y_0 + d \sin \alpha$ in $f(x, y) = 0$, equating to zero the coefficient of the first power of d, and solving for $\tan \alpha$. This value of $\tan \alpha$ makes

$$y - y_0 = \tan \alpha (x - x_0)$$

the equation of the tangent to $f(x, y) = 0$ at (x_0, y_0).

8. If the curve $f(x, y) = 0$ has infinite branches, the values of m and n found by substituting $mx + n$ for y in the equation

$f(x, y) = 0$, and equating to zero the coefficients of the two highest powers of x in the resulting equation, determine the line $y = mx + n$ which meets the curve at two points at infinity; that is, the asymptote.

9. To examine the locus of $y = f(x)$ for maximum and minimum ordinates form $f(x \pm h) - f(x)$. Equate to zero the coefficient of the first power of h, and solve for x. The values of x which make the coefficient of the second power of h positive correspond to minimum, those which make this coefficient negative correspond to maximum ordinates.

10. To determine the points of inflection of $f(x, y) = 0$, substitute $x = x_0 + d \cos \alpha$, $y = y_0 + d \sin \alpha$ in $f(x, y) = 0$. In the resulting equation place the coefficients of d and d^2 equal to zero, and equate the values of $\tan \alpha$ obtained from these equations. The resulting equation, together with the equation of the curve, determines the points of inflection.

Problems. — Plot the numerical algebraic equations:

1. $2x + 3y = 7$.
2. $\frac{x}{4} + \frac{y}{3} = 1$.
3. $xy = 4$.
4. $(x - 2)y = 5$.
5. $(x - 2)(y + 2) = 7$.
6. $x^2 + y^2 = 25$.
7. $x^2 - y^2 = 25$.
8. $x^2 - y^2 = -25$.
9. $y^2 = 10 x$.
10. $y^2 = -10 x$.
11. $4 x^2 + 9 y^2 = 36$.
12. $4 x^2 - 9 y^2 = 36$.
13. $4 x^2 - 9 y^2 = -36$.
14. $y^2 = 10 x - x^2$.
15. $y^2 = x^2 - 10 x$.
16. $x^2 + 10 xy + y^2 = 25$.
17. $x^2 + 10 xy + y^2 + 25 = 0$.
18. $y^2 = 8 x^2 - x^3 + 7$.
19. $x^2 + 2 xy + y^2 = 25$.
20. $x^2 + 10 xy + y^2 = 0$.
21. $y^2 = x^4 - x^2$.
22. $y^2 = x^3 - x^4$.
23. $y^2 = x^3 - x^4$.
24. $y = (x + 2)(x - 3)$.
25. $y = x^2 - 2x - 8$.
26. $y = x^2 - 4x + 4$.
27. $y^2 = (x + 2)(x - 3)$.
28. $y^2 = x^2 - 2x - 8$.
29. $y^2 = x^2 - 4x + 4$.
30. $y = (x - 1)(x - 2)(x - 3)$.
31. $y^2 = x^3 - 6 x^2 + 11 x - 6$.
32. $y = x^4 - 5 x^2 + 4$.
33. $y^2 = x^4 - 5 x^2 + 4$.
34. $y = x^4 + 2 x^3 - 3 x^2 - 4 x + 4$.
35. $y = \frac{x}{1 - x^2}$.

36. $y = \dfrac{2x - 7}{3x + 5}$.

37. $y = \dfrac{y + 5}{3 - x}$.

38. $y = \dfrac{4 - 3x}{5x - 6}$.

39. $y^2 - 2xy - 2 = 0$.

40. $y^2 + 2xy + 3x^2 - 4x = 0$.

41. $y^2 = x^3 - 2x^2 - 8x$.

42. $y = x^3 - 9x^2 + 24x + 3$.

43. $y = (3x - 5)(2x + 9)$.

44. $y^3 = x^3 - 2x^2$.

45. $y^2 + 2xy - 3x^2 + 4x = 0$.

46. $y^2 - 2xy + x^2 + x = 0$.

47. $y^2 + 4xy + 4x^2 - 4 = 0$.

48. $y^2 - 2xy + 2x^2 - 2x = 0$.

49. $y^2 - 2xy + 2x^2 + 2y + x + 3 = 0$.

50. $y^2 - 2xy + x^2 - 4y + x + 4 = 0$.

Plot the real roots of the following equations:

1. $x^2 + 2x - 15 = 0$.
2. $x^3 - 3x - 10 = 0$.
3. $x^2 - 4x + 4 = 0$.
4. $x^2 - 5x + 9 = 0$.
5. $x^3 - 7x + 7 = 0$.
6. $x^3 - 7x - 7 = 0$.
7. $x^3 + 7x + 7 = 0$.
8. $x^3 - 5x + 2 = 0$.
9. $x^4 - 5x^2 + 4 = 0$.
10. $x^4 + x^3 + x^2 + x + 1 = 0$.
11. $x^3 + x^2 + x + 1 = 0$.
12. $x^3 - x^2 + x + 1 = 0$.

Plot the real roots of the following pairs of simultaneous equations:

1. $y^2 = 10x$, $x^2 + y^2 = 25$. Plot these equations to the same axes. The coordinates of the points of intersection of the loci are the pairs of real values of x and y which satisfy each of the given equations. The points of intersection are $(2.07, 4.42)$, $(2.07, -4.42)$.

By the angle of intersection between two curves is meant the angle between the tangents to the curves at their intersection. Hence the angle between two curves is the difference between the angles the tangents to the curves at their intersection make with the X-axis. Calling the angles the tangents to $y^2 = 10x$ and $x^2 + y^2 = 25$ at the point of intersection (x_0, y_0) make with the X-axis a' and a respectively,

$$\tan a' = \frac{5}{y_0}, \quad \tan a = -\frac{x_0}{y_0}.$$

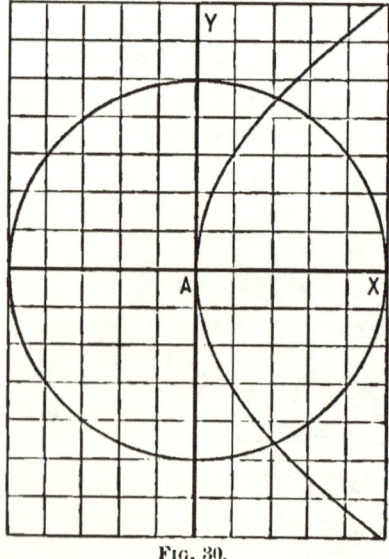

Fig. 30.

Evaluating for $x_0 = 2.07$, $y_0 = 4.48$, $\tan a' = 1.13$, $\tan a = -.47$; whence $a' = 48° 4'$, $a = 154° 50'$, and the angle between the curves is $104° 52'$.

2. $2x + 3y = 5$, $y = \frac{1}{2}x + 3$.
3. $y = 3x + 5$, $x^2 + y^2 = 25$.
4. $x^2 + y^2 = 9$, $y^2 = 10x - x^2$.
5. $y^2 = 10x$, $4x^2 - 9y^2 = 36$.
6. $2x^2 - y^2 = 14$, $x^2 + y^2 = 9$.
7. $y = x^3 - 7x + 7$, $y - x = 0$.
8. $x^2 + y^2 = 25$, $y^2 = 10x - x^2$.
9. $3x^2 + 2y^2 = 7$, $y - 2x = 0$.
10. $y^2 = 4x$, $y - x = 0$.
11. $2x^2 - y^2 = 14$, $x^2 + y^2 = 4$.
12. $x^2 + y^2 = 25$, $x^2 - y^2 = 4$.

Solve the following equations graphically:

1. $x^2 - x - 6 = 0$. Plot $y = x^2$ and $y = x + 6$ to the same axes. For the points of intersection of the loci $x^2 = x + 6$; that is, $x^2 - x - 6 = 0$. Hence the abscissas of the points of intersection of $y = x^2$ and $y = x + 6$ are the real roots of $x^2 - x - 6 = 0$. For all quadratic equations,

$$x^2 + ax + b = 0,$$

the curve $y = x^2$ is the same, and the roots are the abscissas of the points of intersection of the straight line $y = -ax - b$ with this curve.

In like manner the real roots of any trinomial equation $x^n + ax + b = 0$ are the abscissas of the points of intersection of $y = x^n$ and $y + ax + b = 0$.

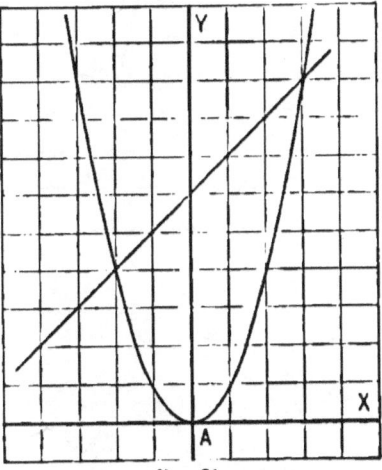

Fig. 31.

2. $x^2 - 3x + 2 = 0$.
3. $x^2 + 5x + 6 = 0$.
4. $x^2 - 4 = 0$.
5. $x^2 - 6x - 10 = 0$.
6. $x^2 - 4x - 15 = 0$.
7. $3x^2 - 12x + 2 = 0$.
8. $x^2 + 5x + 10 = 0$.
9. $x^2 - 5x + 5 = 0$.
10. $x^3 - 7x + 7 = 0$.
11. $x^3 + 7x + 7 = 0$.
12. $x^3 + 7x - 7 = 0$.
13. $x^3 - 7x - 7 = 0$.
14. $x^3 - 10x + 15 = 0$.
15. $x^3 - 10x - 15 = 0$.

Sketch the following literal algebraic equations:

1. $y^2 = x^3 - (b-c)x^2 - bcx$. Here $y = \pm \sqrt{x(x-b)(x+c)}$. Unless numerical values are assigned to b and c, it is impossible to plot the equation by the location of points. However, the general nature of the locus may be determined by discussing the equation. The X-axis is an axis of symmetry, the origin a point of the locus. For $0 < x < b$, y is imaginary; when $x = b$, $y = 0$; for $x > b$, y has two numerically equal real values with opposite signs, increasing indefinitely in numerical value with x. For $0 > x > -c$, y has two numerically equal values with opposite signs; for $x = -c$, $y = 0$; for $x < -c$, y is imaginary. Sketching a curve in accordance with these conditions, a locus of the nature shown in the figure is obtained.

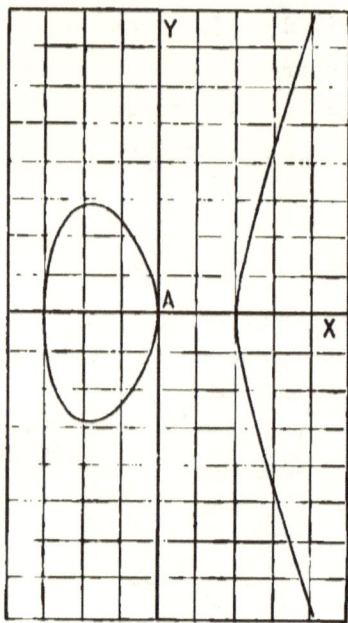

Fig. 32.

2. $\dfrac{x}{a} + \dfrac{y}{b} = 1.$

3. $y = ax.$

4. $y = ax + b.$

5. $xy = m.$

6. $y^2 = 2px.$

7. $x^2 + y^2 = a^2.$

8. $x^2 - y^2 = a^2.$

9. $\dfrac{x^2}{a^2} + \dfrac{y^2}{b^2} = 1.$

10. $\dfrac{x^2}{a^2} - \dfrac{y^2}{b^2} = 1.$

11. $y = (x - a)(x + b).$

12. $y^2 = (x - a)(x + b).$

13. $(x - a)(y - b) = m.$

14. $y = (x - a)(x - b)(x - c).$

15. $y^2 = (x - a)(x - b)(x - c).$

16. $y^2 = (x - a)^2 \dfrac{x - b}{x}.$

17. $y^2 x = 4 a^2 (2a - x).$

18. $y^2 = (x - a)(x + b)(x - c).$

19. $ay^2 = x^3$, the semi-cubic parabola.*

20. $ay = x^3$, the cubic parabola.

21. $y^2 = (x - c_1)(x - c_2)(x - c_3)$, c_1, real, c_2, c_3, conjugate imaginaries.

22. $y^2 = (x - c_1)(x - c_2)(x - c_3)$, c_1, c_2, c_3, real, $c_1 > c_2 > c_3$.

23. $y^2 = (x - c_1)(x - c_2)(x - c_3)$, c_1, c_2, c_3, real, $c_1 = c_2$, $c_1 > c_3$.

24. $y^2 = (x - c_1)(x - c_2)(x - c_3)$, c_1, c_2, c_3, real, $c_1 > c_2$, $c_2 = c_3$.

25. $y^2 = (x - c_1)(x - c_2)(x - c_3)$, $c_1 = c_2 = c_3$.

* The rifling of a cannon, when the bore is rolled out on a plane, technically "developed," is a semi-cubic parabola.

CHAPTER IV

PLOTTING OF TRANSCENDENTAL EQUATIONS

ART. 23. — ELEMENTARY TRANSCENDENTAL FUNCTIONS

Transcendental equations are equations involving transcendental functions.

The elementary transcendental functions are the exponential, logarithmic, circular or trigonometric, and inverse circular functions.

The expression of transcendental functions by means of the fundamental operations of algebra is possible only by means of infinite series.

ART. 24. — EXPONENTIAL AND LOGARITHMIC FUNCTIONS

The general type of the exponential function is $y = b \cdot a^x$, where a is called the base of the exponential function and is always positive.

To plot the exponential function numerically, suppose $b = 1$, $c = 1$, $a = 2$. Then $y = 2^x$ and

$x = -\infty \cdots -4 \quad -3 \quad -2 \quad -1$
$y = \quad 0 \cdots \tfrac{1}{16} \quad \tfrac{1}{8} \quad \tfrac{1}{4} \quad \tfrac{1}{2}$
$\quad\quad 0 \quad 1 \quad 2 \quad 3 \quad 4 \cdots \infty.$
$\quad\quad 1 \quad 2 \quad 4 \quad 8 \quad 16 \cdots \infty.$

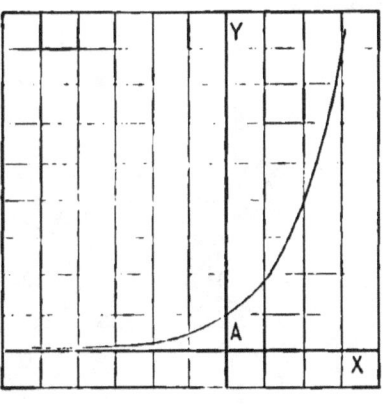

Fig. 33.

For all values of a the locus of $y = a^x$ contains the point $(0, 1)$ and indefinitely approaches the X-axis. Increasing the value of a causes the locus to recede more rapidly from the X-axis for $x > 0$, and to approach the X-axis more rapidly for $x < 0$. When $a = 1$, the locus is a straight line parallel to the X-axis. When $a < 1$, the locus approaches the X-axis for $x > 1$, and recedes from the X-axis for $x < 0$.

When c is not unity the function $y = a^{cx}$ may be written $y = (a^c)^x$, and the base of the exponential function becomes a^c. When b differs from unity, each ordinate of $y = b \cdot a^{cx}$ is the corresponding ordinate of $y = a^{cx}$ multiplied by b.

To plot the exponential function $y = b \cdot a^{cx}$ graphically, compute y_0 and y_1, the values of y corresponding to $x = 0$ and $x = x_1$, where x_1 is any number not zero. Adopt the following notation for corresponding values of x and y:

$$x = \cdots -4x_1 \quad -3x_1 \quad -2x_1 \quad -x_1 \quad 0 \quad x_1 \quad 2x_1 \quad 3x_1 \quad 4x_1 \cdots$$
$$y = \cdots \quad y_{-4} \quad y_{-3} \quad y_{-2} \quad y_{-1} \quad y_0 \quad y_1 \quad y_2 \quad y_3 \quad y_4 \cdots$$

Then $\dfrac{y_{-1}}{y_{-2}} = \dfrac{y_0}{y_{-1}} = \dfrac{y_1}{y_0} = \dfrac{y_2}{y_1} = \dfrac{y_3}{y_2} = \dfrac{y_4}{y_3} = \cdots a^{cx_1}$. On two intersecting straight lines take $OA = y_0$, $OB = y_1$. Join A and B,

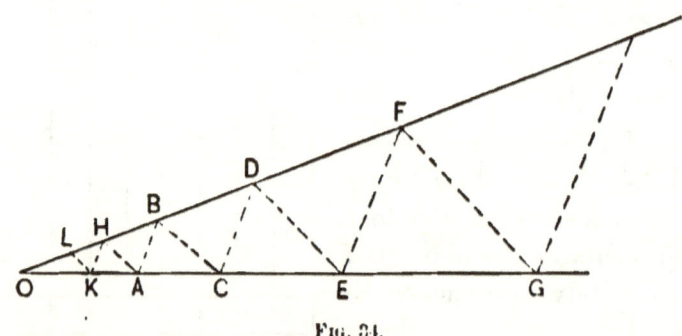

Fig. 34.

draw BC making angle $OBC =$ angle OAB. Then draw CD, DE, EF, $\cdots$, parallel to AB and BC alternately; AH, HK, KL, $\cdots$, parallel to BC and AB alternately. From similar triangles

$$\frac{OK}{OL} = \frac{OH}{OK} = \frac{OA}{OH} = \frac{OB}{OA} = \frac{OC}{OB} = \frac{OD}{OC} = \frac{OE}{OD}.$$

Hence, if $OA = y_0$, $OB = y_1$, it follows that $OL = y_{-3}$, OK

$= y_2$, $OH = y_{-1}$, $OC' = y_2$, $OD = y_3$; that is, the ordinates corresponding to $x = -3x_1, -2x_1, -x_1, 2x_1, 3x_1$ become known and the points of the curve can be located.

The logarithmic function $cx = \log_a(by)$ is equivalent to the exponential function $y = \frac{1}{b}a^{cx}$. When $y = \log x$ is plotted, the logarithm of the product of any two numbers is the sum of the ordinates of the abscissas which represent the numbers, and the product itself is the abscissa corresponding to this sum of the ordinates.

The slide rule is based on this principle. In the slide rule the ordinates of the logarithmic curve are laid off on a straight line from a common point and the ends marked by the corresponding abscissas.

Art. 25.—Circular and Inverse Circular Functions

By definition, $\sin AOP = \frac{PD}{OP} = \frac{P'D'}{OP'}$, and the value of the angle AOP in circular measure is $x = \frac{\text{arc } AP}{OP} = \frac{\text{arc } A'P''}{OP''}$.

Hence, if the radius OA' is the linear unit, the line $P'D'$ is a geometric representative of $\sin AOP$, the arc $A'P''$ a geometric representative of the angle AOP. The measure of the angle AOP is 1 when arc $AP = OP$; that is, the unit of circular measure is the angle at the center which intercepts on the circumference an arc equal to the radius. The unit of circular measure is called the radian.

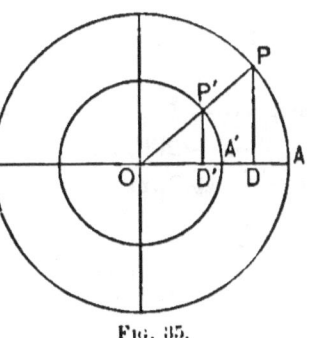

Fig. 35.

The circular measure of four right angles, or $360°$, is $\frac{2\pi OP}{OP}$, or 2π. Hence the radian is equivalent to $\frac{360°}{2\pi} = 57°.3 -$.

ANALYTIC GEOMETRY

Calling angles generated by the anti-clockwise motion of OA positive, angles generated by the clockwise motion of OA negative, there corresponds to every value of the abstract number x a determinate angle.

Unless otherwise specified, angles are expressed in circular measure. When an arc is spoken of without qualification, an arc to radius unity is always understood.

In tables of trigonometric functions angles are generally expressed in degrees. Hence, to plot $y = \sin x$ numerically, assign arbitrary values to x, find the value of the corresponding angle in degrees, and take from the tables the numerical value of $\sin x$.

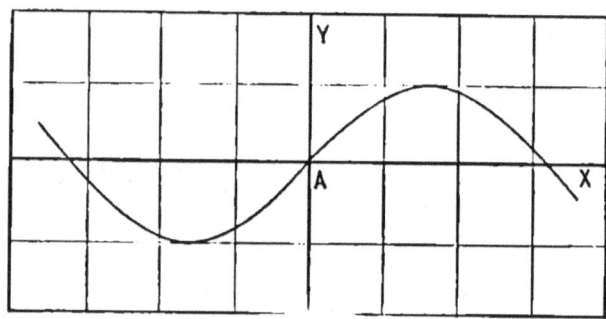

Fig. 36.

$x =$	$\ldots$	$-\frac{7}{2}$	-3	$-\frac{5}{2}$	-2
	$\ldots$	$-200°\,32'$	$-171°\,53'$	$-143°\,14'$	$-114°\,35'$
$y =$	$\ldots$	$+.350$	$-.141$	$-.598$	$-.909$

	$-\frac{3}{2}$	-1	$-\frac{1}{2}$	0
	$-85°\,57'$	$-57°\,18'$	$-28°\,39'$	0
	$-.997$	$-.841$	$-.479$	0

$\frac{1}{2}$	1	$\frac{3}{2}$	2	$\frac{5}{2}$	3	$\ldots$
$28°\,39'$	$57°\,18'$	$85°\,57'$	$114°\,35'$	$143°\,14'$	$171°\,53'$	$\ldots$
$.479$	$.841$	$.997$	$.909$	$.598$	$.141$	$\ldots$

In practical problems the equation frequently occurs in the form $y = \sin(\pi x)$. Here

PLOTTING OF TRANSCENDENTAL EQUATIONS 49

$$x = -2 \quad -\tfrac{7}{4} \quad -\tfrac{3}{2} \quad -\tfrac{5}{4} \quad -1 \quad -\tfrac{3}{4} \quad -\tfrac{1}{2} \quad -\tfrac{1}{4}$$
$$y = 0 \quad \tfrac{1}{2}\sqrt{2} \quad 1 \quad \tfrac{1}{2}\sqrt{2} \quad 0 \quad -\tfrac{1}{2}\sqrt{2} \quad -1 \quad -\tfrac{1}{2}\sqrt{2}$$
$$0 \quad \tfrac{1}{4} \quad \tfrac{1}{2} \quad \tfrac{3}{4} \quad 1 \quad \tfrac{5}{4} \quad \tfrac{3}{2} \quad \tfrac{7}{4} \cdots$$
$$0 \quad \tfrac{1}{2}\sqrt{2} \quad 1 \quad \tfrac{1}{2}\sqrt{2} \quad 0 \quad -\tfrac{1}{2}\sqrt{2} \quad -1 \quad -\tfrac{1}{2}\sqrt{2} \cdots$$

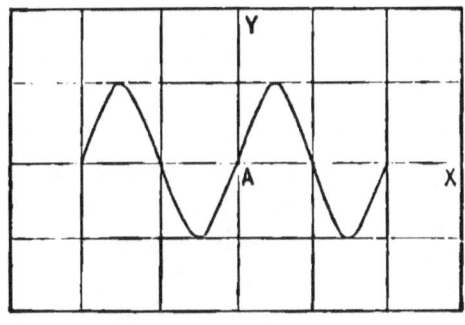

FIG. 87.

To plot $y = \sin x$ graphically, draw a circle with radius unity, divide the circumference into any number of equal parts, and placing the origin of arcs at the origin of coordinates, roll the circle along the X-axis, marking on the X-axis the points of

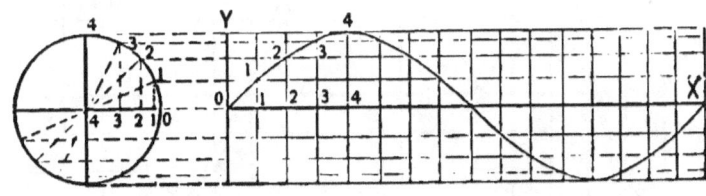

FIG. 38.

division of the circumference 0, 1, 2, 3, 4, 5, 6, ···. Through the points of division of the circumference draw perpendiculars to the diameter through the origin of arcs 00, 11, 22, 33, 44, 55, 66, ···. On the perpendiculars to the X-axis at the points 0, 1, 2, 3, 4, 5, 6, ···, lay off the distances 00, 11, 22, 33, 44, 55, 66, ···, respectively. In this manner any number of points of $y = \sin x$ may be located.

E

On account of the periodicity of $\sin x$, the locus of $y = \sin x$ consists of an infinite number of repetitions of the curve obtained from $x = 0$ to $x = 2\pi$. The locus has maximum ordinates $y = +1$ corresponding to $x = (4n+1)\frac{\pi}{2}$, minimum ordinates $y = -1$ corresponding to $x = (4n+3)\frac{\pi}{2}$, where n is any integer. The locus crosses the x-axis when $x = n\pi$.

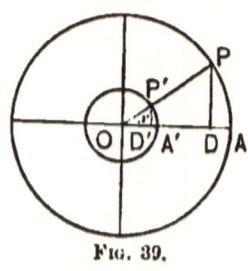

Fig. 39.

To plot $y = 3 \sin x$, it is only necessary to multiply each ordinate of $y = \sin x$ by 3. This is effected graphically by drawing a pair of concentric circles, one with radius unity, the other with radius 3. Since OP'' is the linear unit, $P''D'$ represents $\sin x$, and PD represents $3 \sin x$, while x is represented by the arc $A'P'$.

To plot $y = 3 \sin x + \sin(2x)$, plot $y_1 = 3 \sin x$ and $y_2 = \sin(2x)$ on the same axes. The ordinate of $y = 3 \sin x + \sin(2x)$ cor-

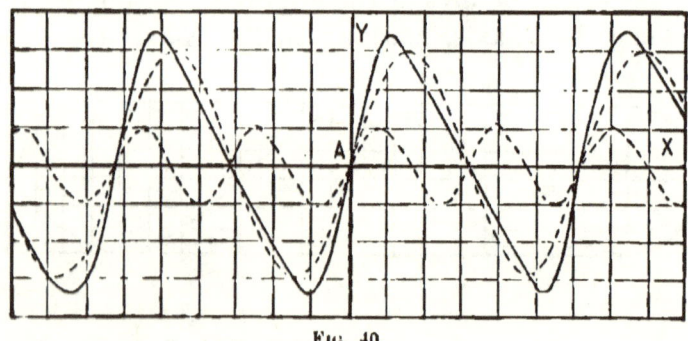

Fig. 40.

responding to any value of x is the sum of the ordinates of $y_1 = 3 \sin x$ and $y_2 = \sin(2x)$ corresponding to the same value of x.

When the sine-function occurs in the form $y = a \sin(\omega t + \theta)$, where ω is uniform angular velocity in radians, t time in seconds, a is called the amplitude, θ the epoch angle. The periodic

PLOTTING OF TRANSCENDENTAL EQUATIONS

time is $t = \dfrac{2\pi}{\omega}$. The construction of the curve is indicated in the figure. The projection of uniform motion in the circumference of a circle on a diameter is called harmonic motion.

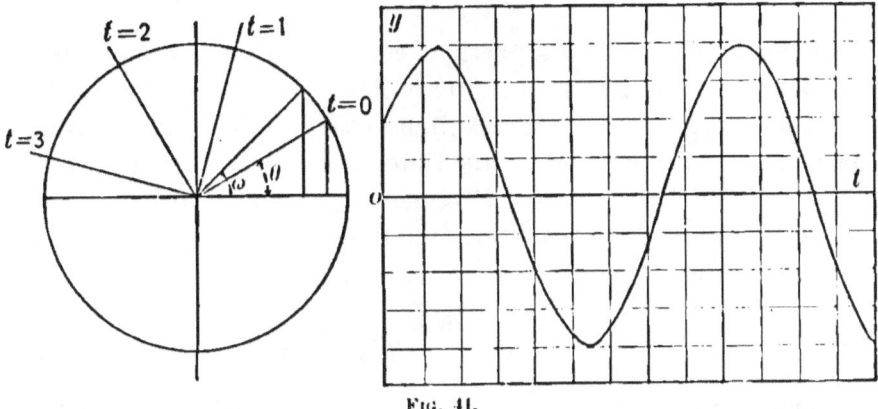

Fig. 41.

To add graphically two sine-functions of equal periods $y_1 = a_1 \sin(\omega t + \theta_1)$, $y_2 = a_2(\sin \omega t + \theta_2)$, draw a pair of concentric circles with radii a_1 and a_2. Let P_1OD and P_2OD be $\omega t + \theta_1$ and $\omega t + \theta_2$ corresponding to the same value of t. Then

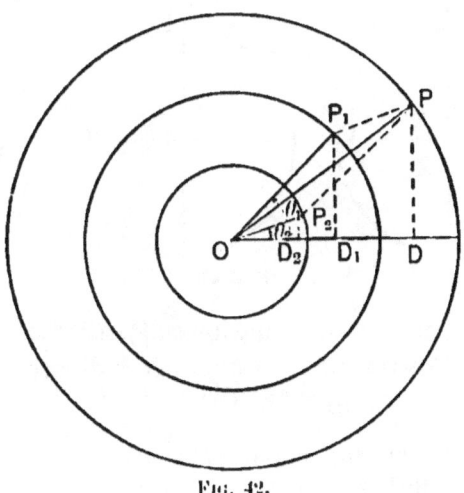

Fig. 42.

$P_1OD - P_2OD = \theta_1 - \theta_2$. The parallelogram on OP_1 and OP_2 for different values of t is the same parallelogram in different positions. This parallelogram has the same angular motion as OP_1 and OP_2. Now $y_1 = P_1D_1$, $y_2 = P_2D_2$, hence

$$PD = P_1D_1 + P_2D_2 = y_1 + y_2,$$

and the sum of the sine-functions corresponding to the circular motions of P_1 and P_2 is the sine-function corresponding to the circular motion of P. The resultant sine-function has the same period as the component sine-functions, its amplitude is OP, its epoch angle the angle POD corresponding to the position of P for $t = 0$. The resultant sine-function is $y = a \sin(\omega t + \theta)$, where $a^2 = a_1^2 + a_2^2 + 2 a_1 a_2 \cos(\theta_1 - \theta_2)$,

$$\tan \theta = \frac{a_1 \sin \theta_1 + a_2 \sin \theta_2}{a_1 \cos \theta_1 + a_2 \cos \theta_2}.*$$

(1) $y = \sin^{-1} x$ is equivalent to $x = \sin y$; (2) $y = 3 \sin^{-1} x$ is equivalent to $x = \sin \frac{y}{3}$; (3) $y = 3 \sin^{-1} \frac{x}{3}$ is equivalent to $x = 3 \sin \frac{y}{3}$; (4) $y = 3 \sin^{-1} \frac{x}{2}$ is equivalent to $x = 2 \sin \frac{y}{3}$. The

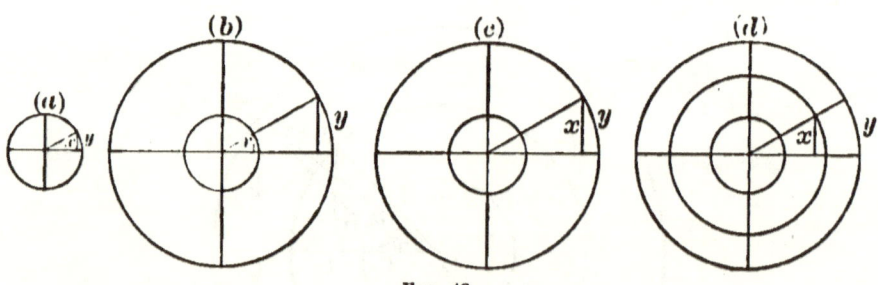

Fig. 43.

graphic interpretation of equations (1), (2), (3), (4) is shown in figures (a), (b), (c), (d), which also indicate the manner of plotting the equations graphically.

* A jointed parallelogram is used for compounding harmonic motions of different periods in Lord Kelvin's tidal clock.

PLOTTING OF TRANSCENDENTAL EQUATIONS   53

The remaining circular and inverse circular functions are plotted in a manner entirely analogous to that employed in plotting $y = \sin x$ and $y = \sin^{-1} x$.

Problems. — Plot 1. $y = 2^x$. 2. $y = 10^x$. 3. $y = (\frac{1}{2})^x$. 4. $y = (.1)^x$.
5. $y = 2^{-x}$. 6. $y = 5 \cdot 2^x$. 7. $y = 3^x$. 8. $y = e^x$.* 9. $y = e^{-x}$. 10. $y = \frac{1}{2}(e^x + e^{-x})$.
This function is called the hyperbolic cosine, and is written $y = \cosh x$.
11. $y = \frac{c}{2}(e^x + e^{-x})$ or $y = c \cosh x$. This is the equation of the catenary,† the form assumed by a perfectly flexible, homogeneous chain whose ends are fastened at two points not in the same vertical.
12. $y = \frac{1}{2}(e^x - e^{-x})$. This is the hyperbolic sine, and is written $\sinh x$.

13. $y = 1^x$.
14. $y = \log_{10} x$.
15. $y = \log_2 x$.
16. $y = \log_{\frac{1}{10}} x$.
17. $y = 3 \log_2 x$.
18. $2x = \log_{10} y$.

19. $x - 2 = \log_{10}(y + 5)$.
20. $x + 5 = \log_{10}(y - 2)$.
21. $x = \log_{10} \frac{2}{y}$.
22. $\frac{x-2}{x+3} = \log_{10}(y+1)$.
23. $y = \sin \frac{x\pi}{2}$.
24. $y = \sin(2x)$.
25. $y = \sin(x + \frac{1}{4}\pi)$.
26. $y = 3 \sin x$.
27. $y = \sin(4x)$.
28. $y = \sin(\frac{1}{2}\pi + 3x)$.
29. $y = 3 + \sin x$.
30. $y = 3 + \sin(\pi + 2x)$.
31. $y = \sin x + 2 \sin \frac{x}{2}$.
32. $y = 2 \sin(2x) + 3 \sin(3x)$.
33. $y = 3 \sin(2 + 2x) + 5 \sin(1 + 4x)$.
34. $y = \cos x$.

35. $y = 5 \cos x$.
36. $y = 2 \cos(1 + 5x)$.
37. $y = \tan x$.
38. $y = 2 + \tan(1 + x)$.
39. $y = \sec x$.
40. $y = 2 \sec x$.
41. $y = \csc x$.
42. $y = \csc(x - 1)$.
43. $y = \text{vers } x$.
44. $y = \text{covers } x$.
45. $x = \sin^{-1} \frac{y}{2}$.

---

* e represents the base of the Napierian system of logarithms, a transcendental number, whose value to nine places is 2.718281828. $y = e^x$ may be plotted graphically by computing the values of y corresponding to any two values of x; numerically by writing the function in the form $x = \log_e y$ and using a table of Napierian logarithms.

† The catenary was invented by John and James Bernoulli. The center of gravity of the catenary is lower than for any other position of the same chain with the same fixed points.

46. $2y = \cos^{-1} x$.
47. $y = \frac{1}{2}\tan^{-1} x$.
48. $y = \sec^{-1}(x-3)$.
49. $y = 2 + \text{vers}^{-1} x$.
50. $y = 3\cos^{-1}\frac{x}{3}$.
51. $y = 5\sin^{-1}\frac{x}{4}$.
52. $x - 2 = \sin^{-1} y$.
53. $x + 3 = \sin^{-1}(y-2)$.
54. $x = \cos^{-1}(y-1)$.
55. $y + 2 = \cos^{-1}(x-2)$.
56. $y = \sin(\frac{1}{4}\pi t)$.
57. $y = \sin(\frac{1}{4}\pi t + \frac{1}{2}\pi)$.
58. $y = \sin(\frac{1}{2}\pi t + \frac{1}{2}\pi) + \sin(\frac{1}{2}\pi t + \pi)$.
59. $y = \sin(\frac{1}{4}\pi t) + \sin(\frac{1}{4}\pi t + \frac{1}{2}\pi)$.
60. $y = \sin(\frac{1}{4}\pi t + \frac{1}{8}\pi) + \sin(\frac{1}{4}\pi t + \frac{1}{2}\pi)$.

The elementary transcendental functions are of great importance in mathematical physics. For instance, if a steady electric current I flows through a circuit, the strength i of the current t seconds after the removal of the electromotive force is given by the exponential function $i = Ie^{-\frac{Rt}{L}}$, where R and L are constants of the circuit.

The quantity of light that penetrates different thicknesses of glass is a logarithmic function of the thickness.

The sine-function is the element by whose composition any single-valued periodic function may be formed. Vibratory motion and wave motion are periodic. The sine-function or, as it is also called, the simple harmonic function, thus becomes of fundamental importance in the mathematical treatment of heat, light, sound, and electricity.

Art. 26. — Cycloids

A circle rolls along a fixed straight line. The curve traced by a point fixed in the circumference of the circle is called a cycloid.* The fixed line is called the base, the point whose distance from the base is the diameter of the generating circle the vertex, the perpendicular from the vertex to the base the

* Curves generated by a point fixed in the plane of a curve which rolls along some fixed curve are called by the general name "roulettes." Cycloids are a special class of roulettes.

PLOTTING OF TRANSCENDENTAL EQUATIONS

axis of the cycloid. The coordinates of any point of the cycloid may be expressed as transcendental functions of a variable angle.

Take the base of the cycloid as X-axis, the perpendicular to the base where the cycloid meets the base as Y-axis, and call the angle made by the radius of the generating circle to the tracing point with the vertical diameter θ. By the nature of the

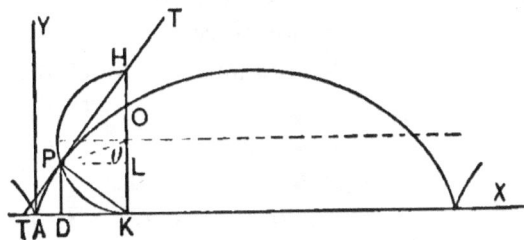

Fig. 44.

cycloid $AK =$ arc $PK = r\theta$, $y = PD = OK - OL = r - r\cos\theta$, $x = AD = AK - DK = r\theta - r\sin\theta$. Hence (1) $x = r\theta - r\sin\theta$, $y = r - r\cos\theta$ for every value of θ determine a point of the cycloid. The equation of the cycloid between x and y is obtained either directly from the figure,

$$x = AK - DK = \text{arc } PK - PL = r\text{ vers}^{-1}\frac{y}{r} - \sqrt{2ry - y^2};$$

or by eliminating θ between equations (1),

$$\theta = \cos^{-1}\left(1 - \frac{y}{r}\right) = \text{vers}^{-1}\frac{y}{r}, \quad \sin\theta = \frac{1}{r}\sqrt{2ry - y^2},$$

hence
$$x = r\text{ vers}^{-1}\frac{y}{r} - \sqrt{2ry - y^2}.$$

Now $r\text{ vers}^{-1}\frac{y}{r}$ has for the same value of y an infinite number of values differing by 2π, and $\sqrt{2ry - y^2}$ is a two-valued function which is real only for values of y from 0 to $+2r$. Hence the equation determines an infinite number of values of x for every value of y between 0 and $+2r$. This agrees with the nature of the curve as determined by its generation.

By observing that the center of the generating circle is always in the line parallel to the base at a distance equal to the radius of the generating circle, the generating circle may readily be placed in position for locating any point of the cycloid. At the instant the point P is being located the generating circle is revolving about K, hence the generating point P tends at that instant to move in the circumference of a circle whose center is K and radius the chord KP. The tangent to the cycloid at P is therefore the perpendicular to the chord KP at P, that is the tangent is the chord PH of the generating circle.*

The perpendicular to the tangent to a curve at the point of tangency is called the normal to the curve at that point. Hence the chord KP is the normal to the cycloid at P.

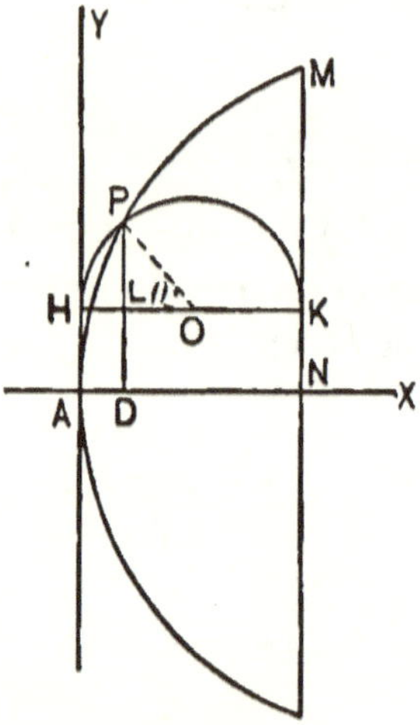

Take the axis of the cycloid as X-axis, the tangent at the vertex as Y-axis. By the nature of the cycloid

$$MK = \text{arc } PK,$$
$$MN = \text{semi-circumference } KPH.$$
$$x = AD = HL = OH - OL = r - r \cos \theta,$$
$$y = PD = LD + PL = (MN - MK) + PL$$
$$= \text{arc } HP + PL = r\theta + r \sin \theta.$$

Fig. 45.

That is,

$$(1) \quad x = r - r \cos \theta,$$
$$y = r\theta + r \sin \theta,$$

determine for every value of θ a point of the cycloid. The equation between x and y is found either directly from the figure,

$$y = LD + PL = r \text{ vers}^{-1} \frac{x}{r} + \sqrt{2rx - x^2};$$

* This method of drawing a tangent to the cycloid is due to Descartes.

PLOTTING OF TRANSCENDENTAL EQUATIONS

or by eliminating θ between equations (1),

$$\theta = \cos^{-1}\left(1 - \frac{x}{r}\right) = \mathrm{vers}^{-1}\frac{x}{r},$$

$$\sin\theta = \frac{1}{r}\sqrt{2rx - x^2},$$

hence
$$y = r\,\mathrm{vers}^{-1}\frac{x}{r} + \sqrt{2rx - x^2}.*$$

Art. 27. — Prolate and Curtate Cycloids

When the generating point instead of being on the circumference is a point fixed in the plane of the rolling circle, the curve generated is called the prolate cycloid when the point is within the circumference, the curtate cycloid when the point is without the circumference. Let a be the distance from the center to the generating point. From the figures the equations of these curves are readily seen to be

$$x = r\theta - a\sin\theta, \quad y = r - a\cos\theta.$$

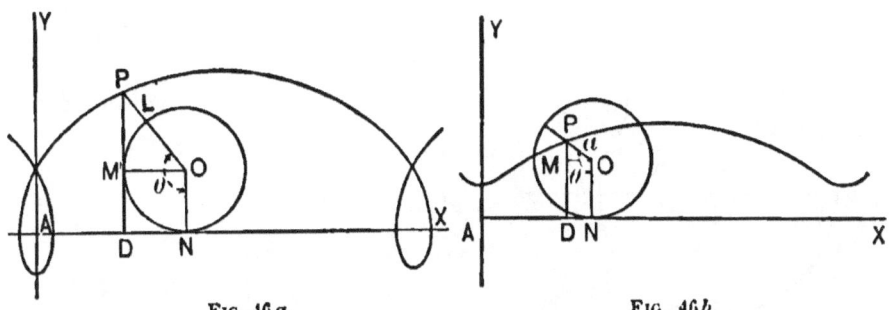

Fig. 46 a.   Fig. 46 b.

* If the cycloid is concave up and the tangent at the vertex horizontal, the time required by a particle sliding down the cycloid, supposed frictionless, to reach the vertex is independent of the starting point. On account of this property, discovered by Huygens in 1673, the cycloid is called the tautochrone. The frictionless curve along which a body must slide to pass from one point to another in the shortest time is a cycloid. On account of this property, discovered by John Bernoulli in 1696, the cycloid is called the brachistochrone.

Art. 28. — Epicycloids and Hypocycloids

If a circle rolls along the circumference of a fixed circle, the curve generated by a point fixed in the circumference of the rolling circle is called an epicycloid if the circle rolls along the outside, an hypocycloid if the circle rolls along the inside of the circumference of the fixed circle. By the nature of the epicycloid arc HO = arc HP, that is $R \cdot \theta = r \cdot \phi$. From the figure $x = AD = AL + DL$

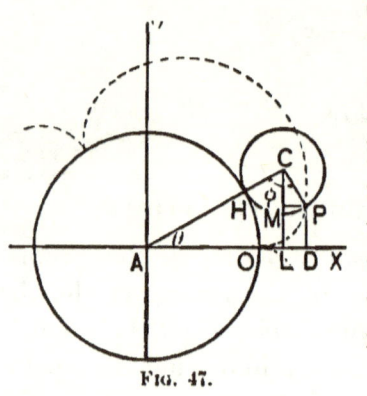

Fig. 47.

$$= (R + r)\cos \theta + r \cos CPM.$$

Now $CPM = 180° - (\phi + \theta) = 180° - \left(\dfrac{R}{r}\theta + \theta\right) = 180° - \dfrac{R+r}{r}\theta.$

Hence $x = (R + r)\cos \theta - r \cos \dfrac{R+r}{r}\theta.$ $y = PD = CL - CM$

$$= (R + r)\sin \theta - r \sin \dfrac{R+r}{r}\theta.$$

By the nature of the hypocycloid $R \cdot \theta = r \cdot \phi$, hence $\phi = \dfrac{R}{r} \cdot \theta.$ $x = AD$

$$= AL - PM$$
$$= (R - r)\cos \theta - r \cos CPM.$$

Now $CPM = 180° - \phi + \theta$

$$= 180° - \dfrac{R-r}{r}\theta.$$ Hence

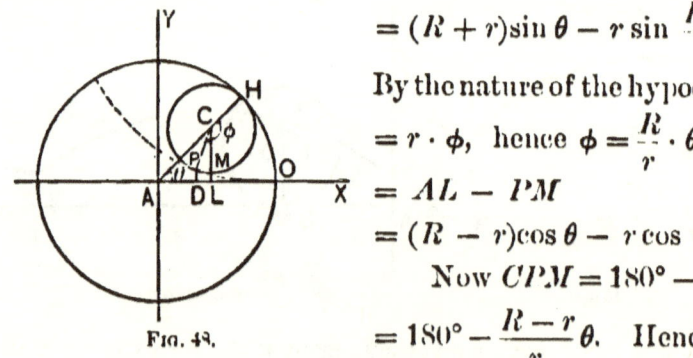

Fig. 48.

$x = (R - r)\cos \theta + r \cos \dfrac{R-r}{r}\theta.$

$y = PD = CL - CM = (R - r)\sin \theta - r \sin \dfrac{R-r}{r}\theta.$*

---

* Epicycloids and hypocycloids are used in constructing gear teeth.

Art. 29. — Involute of Circle

A string whose length is the circumference of a circle is wound about the circumference. One end is fastened at O and the string unwound. If the string is kept stretched, its free

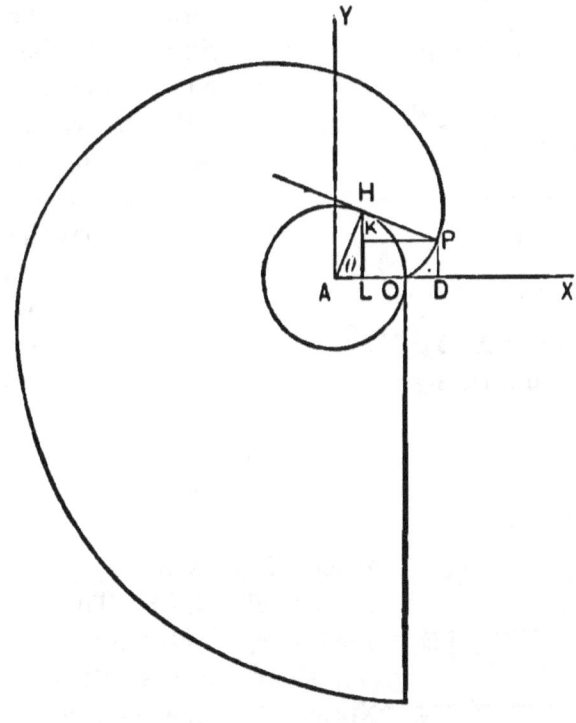

Fig. 49.

end traces a curve which is called the involute of the circle. From the nature of the involute, HP is tangent to the fixed circle and equals the arc HO, which equals $R\theta$.

$$x = AD = AL + KP = R \cos \theta + R\theta \sin \theta,$$
$$y = PD = HL - HK = R \sin \theta - R\theta \cos \theta.*$$

* The involute is also used in constructing gear teeth.

CHAPTER V

TRANSFORMATION OF COORDINATES

Art. 30. — Transformation to Parallel Axes

Let P be any point in the plane. Referred to the axes X, Y the point P is represented by (x, y); referred to the parallel axes X_1, Y_1, the point P is represented by (x_1, y_1). Let the origin A_1 be (m, n) when referred to the axes X, Y.

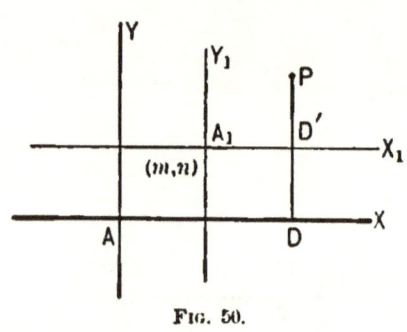

Fig. 50.

From the figure $x = m + x_1$, $y = n + y_1$. Since (x, y) and (x_1, y_1) represent the same point P, if $f(x, y) = 0$ is the equation of a certain geometric figure when interpreted with reference to the axes X, Y, $f(m + x_1, n + y_1) = 0$ is the equation of the same geometric figure when interpreted with reference to the axes X_1, Y_1.

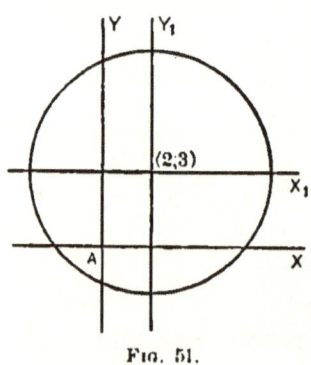

Fig. 51.

EXAMPLE. — The equation of the circle whose radius is 5, center $(2, 3)$ is (1) $(x-2)^2 + (y-3)^2 = 25$. Draw a set of axes X_1, Y_1 parallel to X, Y through $(2, 3)$. Then $x = 2 + x_1$, $y = 3 + y_1$. Substituting in equation (1), there results (2) $x_1^2 + y_1^2 = 25$. Notice that the equation of a geometric figure depends on the position of the geometric figure with respect to the axes.

Art. 31.—From Rectangular Axes to Rectangular

Let (x, y) represent any point in the plane referred to the axes X, Y; (x_1, y_1) the same point referred to the axes X_1, Y_1, where X_1, Y_1 are obtained by turning X, Y about A through the angle α. Now

$$x = AD = AH - KD'$$
$$= x_1 \cos \alpha - y_1 \sin \alpha,$$
$$y = PD = D'H + PK$$
$$= x_1 \sin \alpha + y_1 \cos \alpha.$$

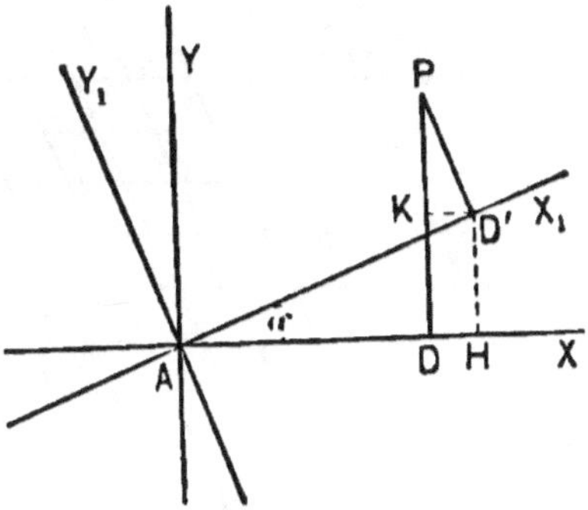

Since (x, y) and (x_1, y_1) represent the same point P, $f(x, y) = 0$ interpreted on the X, Y axes and

Fig. 52.

$$f(x_1 \cos \alpha - y_1 \sin \alpha, \; x_1 \sin \alpha + y_1 \cos \alpha) = 0$$

interpreted on the X_1, Y_1 axes represent the same geometric figure.

Example.—$y = x + 4$ is the equation of a straight line. To find a set of rectangular axes, the origin remaining the same, to which when this line is referred its equation takes the form $y_1 = n$, substitute in the given equation

$$x = x_1 \cos \alpha - y_1 \sin \alpha,$$
$$y = x_1 \sin \alpha + y_1 \cos \alpha.$$

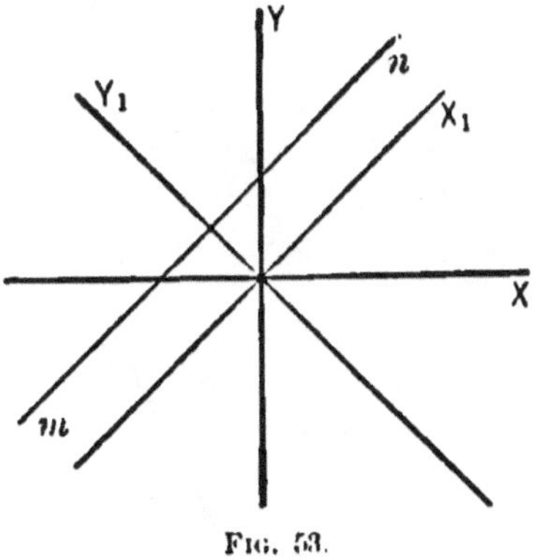

There results

$$x_1 (\sin \alpha - \cos \alpha) + y_1 (\sin \alpha + \cos \alpha) = 4,$$

and this equation takes the required form when

$$\sin \alpha - \cos \alpha = 0,$$

Fig. 53.

that is, when $\alpha = 45°$. Substituting this value of α, the transformed equation becomes $y_1 = 2\sqrt{2}$.

Art. 32. — Oblique Axes

Hitherto the axes of reference have been perpendicular to each other. The position of a point in the plane can be equally well determined when the axes are oblique. The ordinate of the point P referred to the oblique axes X, Y is the distance and direction of the point from the X-axis, the distance being measured on a parallel to the Y-axis, the side of the X-axis on which the point lies being indicated by the algebraic sign prefixed to the number expressing this distance.

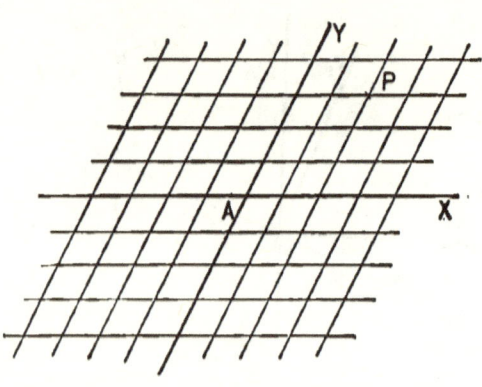

Fig. 54.

Similarly the abscissa of the point P is the distance and direction of the point from the Y-axis, the distance being measured on a parallel to the X-axis, the algebraic sign prefixed to this distance denoting on what side of the Y-axis the point lies.

Problems. — The angle between the oblique axes being $45°$:

1. Locate the points $(3, -2)$; $(-5, 4)$; $(0, 8)$; $(-4, -7)$; $(2\frac{1}{3}, -3)$; $(\sqrt{5}, -\sqrt{7})$; $(-2\frac{1}{2}, \sqrt{10})$.

2. Plot $y = 3x$; $y = 3x + 5$; $y = -2x + 3$.

3. Plot $x^2 + y^2 = 16$; $y^2 = 4x$; $\frac{x^2}{9} + \frac{y^2}{4} = 1$.

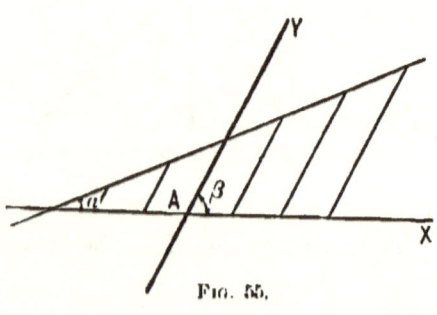

Fig. 55.

Observe that the geometric figure represented by an equation depends on the system of coordinates used in plotting the equation.

4. Find the equation of a straight line referred to oblique axes including an angle β. The method used to find the equation of a straight line referred to

rectangular axes shows that the equation of a straight line referred to oblique axes is $y = mx + n$, where $m = \dfrac{\sin \alpha}{\sin(\beta - \alpha)}$, and n is the intercept of the line on the Y-axis.

5. Show that $\sqrt{(x' - x'')^2 + (y' - y'')^2 + 2(x' - x'')(y' - y'')\cos\beta}$ is the distance between the points (x', y'), (x'', y'') when the angle between the axes is β.

6. Find the equation of the circle whose radius is R, center (m, n), when the angle between the axes is β.

7. Show that double the area of the triangle whose vertices are

$$(x_1, y_1), (x_2, y_2), (x_3, y_3)$$

is $\quad \{y_1(x_3 - x_2) + y_2(x_1 - x_3) + y_3(x_2 - x_1)\}\sin\beta.$

ART. 33. — FROM RECTANGULAR AXES TO OBLIQUE

It is sometimes desirable to find the equation of a geometric figure referred to oblique axes when the equation of this figure referred to rectangular axes is known. This manner of obtaining the equation of a figure referred to oblique axes is frequently a simpler problem than to obtain the equation directly. To accomplish the transformation, the rectangular coordinates of a point must be expressed in terms of the oblique coordinates of the same point. From the figure

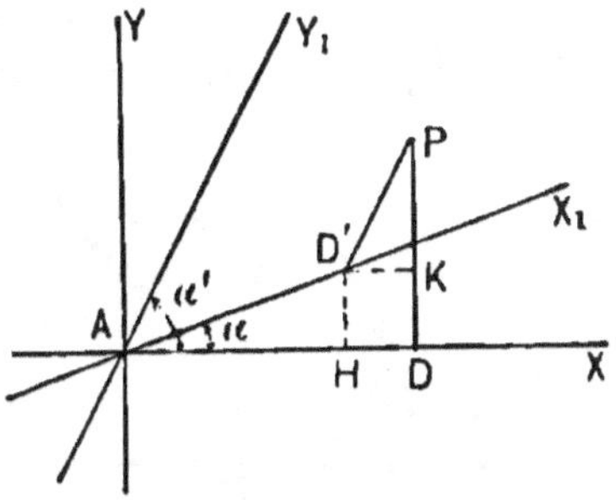

FIG. 56.

$$x = AD = AH + D'K = x_1 \cos\alpha + y_1 \cos\alpha',$$
$$y = PD = D'H + PK = x_1 \sin\alpha + y_1 \sin\alpha'.$$

EXAMPLE. — To find the equation of the hyperbola referred to its asymptotes from the common equation of the hyperbola, $\dfrac{x^2}{a^2} - \dfrac{y^2}{b^2} = 1.$

The asymptotes of the hyperbola, $y = \pm \frac{b}{a} x$, are the diagonals of the rectangle on the axes. Hence

$$\sin \alpha = -\frac{b}{\sqrt{a^2 + b^2}}, \qquad \cos \alpha = \frac{a}{\sqrt{a^2 + b^2}},$$

$$\sin \alpha' = \frac{b}{\sqrt{a^2 + b^2}}, \qquad \cos \alpha' = \frac{a}{\sqrt{a^2 + b^2}},$$

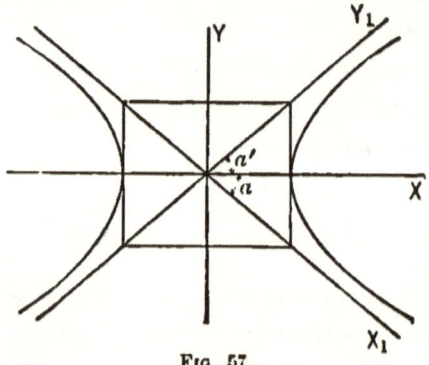

Fig. 57.

and the transformation formulas become

$$x = \frac{a}{\sqrt{a^2 + b^2}} (x_1 + y_1),$$

$$y = \frac{b}{\sqrt{a^2 + b^2}} (y_1 - x_1).$$

Substituting in the common equation of the hyperbola and reducing, $x_1 y_1 = \frac{a^2 + b^2}{4}$, the equation of the hyperbola referred to its asymptotes.

The formulas for passing from oblique axes to rectangular, the origin remaining the same, are

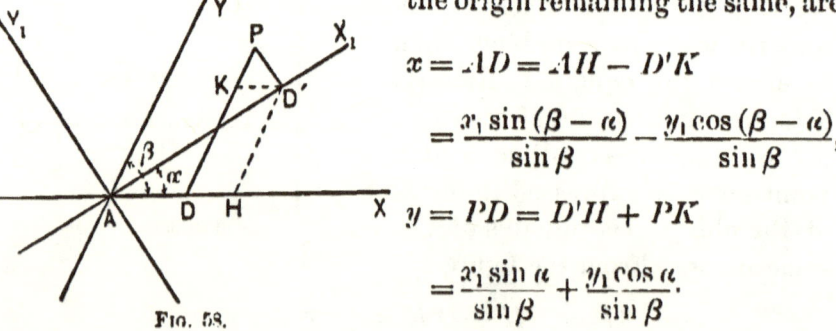

Fig. 58.

$$x = AD = AH - D'K$$

$$= \frac{x_1 \sin (\beta - \alpha)}{\sin \beta} - \frac{y_1 \cos (\beta - \alpha)}{\sin \beta},$$

$$y = PD = D'H + PK$$

$$= \frac{x_1 \sin \alpha}{\sin \beta} + \frac{y_1 \cos \alpha}{\sin \beta}.$$

Art. 34. — General Transformation

The general formulas for transforming from one set of rectilinear axes to another set of rectilinear axes, the origin of the second set when referred to the first set being (m, n), are

$$x = AD = AR + A_1T + D'K,$$
$$= m + \frac{x_1 \sin(\beta - \alpha)}{\sin \beta} + \frac{y_1 \sin(\beta - \alpha')}{\sin \beta},$$
$$y = PD = A_1R + D'T + PK,$$
$$= n + \frac{x_1 \sin \alpha}{\sin \beta} + \frac{y_1 \sin \alpha'}{\sin \beta}.$$

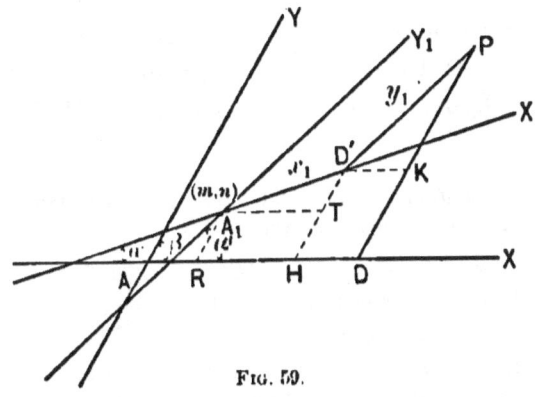

Fig. 59.

From these general formulas all the preceding formulas may be derived by substituting for m, n, β, α, α' their values in each special case. However, if it is observed that in every case the figure used in deriving the transformation formulas is constructed by drawing the coordinates of any point P referred to the original axes, and the coordinates of the same point referred to the new axes, then through the foot of the new ordinate parallels to the original axes, it is simpler to derive these formulas directly from the figure, whenever they are needed.

Art. 35. — The Problem of Transformation

An examination of the transformation formulas shows that the values of the rectilinear coordinates of any point in terms of any other rectilinear coordinates of the same point are of the first degree. Hence transformation from one set of recti-

F

linear coordinates to another rectilinear set does not change the degree of the equation of the geometric figure.

Two classes of problems are solved by the transformation of coordinates:

I. Having given the equation of a geometric figure referred to one set of axes, to find the equation of the same geometric figure referred to another set of axes.

II. Having given the equation of a geometric figure referred to one set of axes, to find a second set of axes to which when the geometric figure is referred its equation takes a required form.

Problems. — Transform to parallel axes, given the coordinates of the new origin referred to the original axes.

1. $y - 2 = 3(x + 5)$, origin $(-5, 2)$.
2. $(x - 3)(y - 4) = 5$, origin $(3, 4)$.
3. $y = 2x + 5$, origin $(0, 5)$.
4. $x^2 + y^2 + 2x + 4y = 4$, origin $(-1, -2)$.
5. $x^2 + y^2 + 6y = 7$, origin $(0, -3)$.
6. $x^2 + y^2 - 6x = 16$, origin $(3, 0)$.
7. $y^2 + 4y - 6x = 4$, origin $(0, -2)$.
8. $25(y + 4)^2 + 16(x - 5)^2 = 400$, origin $(5, -4)$.
9. $x^2 + y^2 = 25$, origin $(-5, 0)$.
10. $x^2 + y^2 = 25$, origin $(0, -5)$.
11. $x^2 + y^2 = 25$, origin $(-5, -5)$.
12. $\frac{x^2}{a^2} + \frac{y^2}{b^2} = 1$, origin $(-a, 0)$.
13. $\frac{x^2}{a^2} + \frac{y^2}{b^2} = 1$, origin $(0, -b)$.
14. $\frac{x^2}{a^2} + \frac{y^2}{b^2} = 1$, origin $(-a, -b)$.
15. $\frac{x^2}{a^2} - \frac{y^2}{b^2} = 1$, origin $(a, 0)$.

Transform from one rectangular set to a second rectangular set, the second set being obtained by turning the first about the origin through $45°$.

16. $x^2 + y^2 = 4$.
17. $x^2 - y^2 = 4$.
18. $y + x = 5$.
19. $y^2 + xy - x^2 = 6$.
20. $y^2 + 4xy + x^2 = 8$.
21. $y^3 - 3xy + x^3 = 0$.
22. $y^3 + 3xy - x^3 = 0$.

Notice that to plot equations 21 and 22 directly requires the solution of a cubic equation, whereas the transformed equations are plotted by the solution of a quadratic equation.

TRANSFORMATION OF COORDINATES

In the following problems the first equation is the equation of a geometric figure referred to rectangular axes. The origin of a parallel set of axes is to be found to which when the geometric figure is referred its equation is the second equation given.

23. $y + 2 = 4(x - 3)$; $y = 4x$.   26. $y^2 - x^2 - 10x = 0$; $x^2 - y^2 = 25$.
24. $(x + 1)(y + 5) = 4$; $xy = 4$.   27. $y^2 - 10(x + 5) = 0$; $y^2 = 10x$.
25. $y^2 + x^2 + 10x = 0$; $x^2 + y^2 = 25$.   28. $y^2 + x^2 + 4y - 2x = 11$; $x^2 + y^2 = 16$.

In the following problems the first equation is the equation of a geometric figure referred to rectangular axes; find the inclination of a second set of rectangular axes to the given, origin remaining the same, to which when the geometric figure is referred its equation is the second equation given.

29. $y = x + 4$; $y = 2\sqrt{2}$.
30. $x^2 + y^2 = 25$; $x^2 + y^2 = 25$.
31. $x^2 - y^2 = 1$; $xy = \frac{1}{2}$.
32. $\dfrac{x^2}{a^2} - \dfrac{y^2}{b^2} = 1$; $xy = \dfrac{a^2 + b^2}{4}$.
33. $y^3 - 3axy + x^3 = 0$; $y^2 = \dfrac{3\sqrt{2}\,ax^2 - 2x^3}{2x + 3\sqrt{2}\,a}$.

Construct the locus of the first equation in the following problems by drawing the axes X_1, Y_1 and plotting the second equation.

34. $y - 2 = \log(x + 3)$; $y_1 = \log x_1$.   36. $y + 3 = 2^{x+4}$; $y_1 = 2^{x_1}$.
35. $y = 3\sin(x + 5)$; $y_1 = 3\sin x_1$.   37. $y + 5 = \tan(x - 3)$; $y_1 = \tan x_1$.

38. From the common equation of the hyperbola, $\dfrac{x^2}{a^2} - \dfrac{y^2}{b^2} = 1$, obtain the equation of the hyperbola referred to oblique axes through the center, such that $\tan a \tan a' = \dfrac{b^2}{a^2}$.

The transformation formulas are $x = x_1 \cos a + y_1 \cos a'$,
$y = x_1 \sin a + y_1 \sin a'$;
the transformed equation

$\left(\dfrac{\cos^2 a}{a^2} - \dfrac{\sin^2 a}{b^2}\right)x_1^2$
$+ 2\left(\dfrac{\cos a \cos a'}{a^2} - \dfrac{\sin a \sin a'}{b^2}\right)x_1 y_1$
$+ \left(\dfrac{\cos^2 a'}{a^2} - \dfrac{\sin^2 a'}{b^2}\right)y_1^2 = 1$.

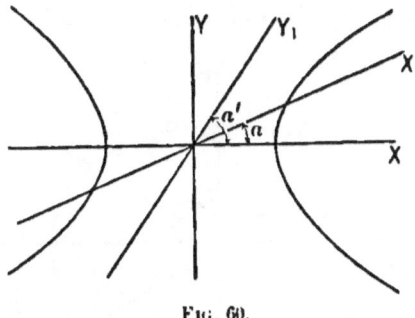

Fig. 60.

The condition $\tan a \tan a' = \dfrac{b^2}{a^2}$ renders the coefficient of $x_1 y_1$ zero, and the equation of the hyperbola referred to the oblique axes becomes

$$\left(\frac{\cos^2 a}{a^2} - \frac{\sin^2 a}{b^2}\right)x_1^2 + \left(\frac{\cos^2 a'}{a^2} - \frac{\sin^2 a'}{b^2}\right)y_1^2 = 1.$$

Since only values of a and a' less than $180°$ need be considered, the condition $\tan a \tan a' = \frac{b^2}{a^2}$ shows that a and a' are either both less than $90°$ or both greater than $90°$, and that if $\tan a < \frac{b}{a}$, $\tan a' > \frac{b}{a}$. Since the equations of the asymptotes of the hyperbola are $y = \pm \frac{b}{a}x$, it follows that if the X_1-axis intersects the hyperbola, the Y_1-axis cannot intersect it. Calling the intercepts of the hyperbola on the X_1- and Y_1-axis respectively a_1 and $b_1\sqrt{-1}$, the equation referred to the oblique axes becomes

$$\frac{x_1^2}{a_1^2} - \frac{y_1^2}{b_1^2} = 1.$$

39. From the common equation of the ellipse, $\frac{x^2}{a^2} + \frac{y^2}{b^2} = 1$, obtain the equation of the ellipse referred to oblique axes such that $\tan a \tan a' = -\frac{b^2}{a^2}$.

40. From the common equation of the parabola $y^2 = 2px$ obtain the equation of the parabola referred to oblique axes, origin (m, n) on the parabola, the X_1-axis parallel to the axis of the parabola, the Y_1-axis tangent to the parabola.

$n^2 = 2pm$, $a = 0$, and, since the equation of the tangent to $y^2 = 2px$ at (m, n) is $y = \frac{p}{n}(x+m)$, $\tan a' = \frac{p}{n}$. The transformation formulas become $x = m + x_1 + y_1 \cos a'$, $y = n + y_1 \sin a'$, and the transformed equation reduces to $y_1^2 = 2\frac{p^2 + n^2}{p}x_1$, or $y_1^2 = 2(p + 2m)x_1$.

41. To determine a set of oblique axes, with the origin at the center, to which when the ellipse is referred, its equation has the same form as the common equation of the ellipse $\frac{x^2}{a^2} + \frac{y^2}{b^2} = 1$.

The substitution of
$$x = x_1 \cos a + y_1 \cos a',$$
$$y = x_1 \sin a + y_1 \sin a'$$
transforms the equation $\frac{x^2}{a^2} + \frac{y^2}{b^2} = 1$ into

$$\left(\frac{\cos^2 a}{a^2} + \frac{\sin^2 a}{b^2}\right)x_1^2$$
$$+ 2\left(\frac{\cos a \cos a'}{a^2} + \frac{\sin a \sin a'}{b^2}\right)x_1 y_1$$
$$+ \left(\frac{\cos^2 a'}{a^2} + \frac{\sin^2 a'}{b^2}\right)y_1^2 = 1.$$

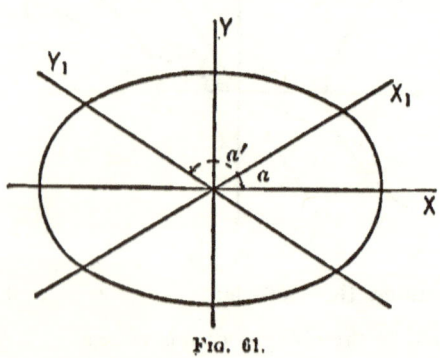

Fig. 61.

The problem requires that the coefficient of $x_1 y_1$ be zero, hence
$$\tan \alpha \tan \alpha' = -\frac{b^2}{a^2}.$$

The problem is indeterminate, since the equation between α and α' admits an infinite number of solutions. Let α and α' in the figure represent one solution, then $\left(\dfrac{\cos^2 \alpha}{a^2} + \dfrac{\sin^2 \alpha}{b^2}\right) x_1^2 + \left(\dfrac{\cos^2 \alpha'}{a^2} + \dfrac{\sin^2 \alpha'}{b^2}\right) y_1^2 = 1$ is the equation of the ellipse referred to the axes X_1, Y_1. Call the intercepts of the ellipse on the axes X_1 and Y_1 respectively a_1 and b_1, and the equation becomes $\dfrac{x_1^2}{a_1^2} + \dfrac{y_1^2}{b_1^2} = 1$.

When the equation of the ellipse referred to a pair of lines through the center contains only the squares of the unknown quantities, these lines are called conjugate diameters of the ellipse. The condition of conjugate diameters of the ellipse is $\tan \alpha \tan \alpha' = -\dfrac{b^2}{a^2}$.

42. Determine a set of oblique axes, with the origin at the center, to which, when the hyperbola is referred, its equation takes the same form as the common equation of the hyperbola.

The result, $\tan \alpha \tan \alpha' = \dfrac{b^2}{a^2}$, shows that the problem is indeterminate. $\tan \alpha \tan \alpha' = \dfrac{b^2}{a^2}$ is the condition of conjugate diameters of the hyperbola.

43. Determine a set of oblique axes, origin at center, to which, when the hyperbola is referred, its equation takes the form $xy = c$.

44. Determine origin and direction of a set of oblique axes to which, when the parabola is referred, its equation has the same form as the common equation of the parabola.

45. Show that the equation of the parabola $y^2 = 2px$ when referred to its focal tangents becomes $x^{\frac{1}{2}} + y^{\frac{1}{2}} = a^{\frac{1}{2}}$, where a is the distance from the new origin to the points of tangency.

CHAPTER VI

POLAR COORDINATES

ART. 36. — POLAR COORDINATES OF A POINT

In the plane, suppose the point A and the straight line AX through A fixed. A is called the pole, AX the polar axis.

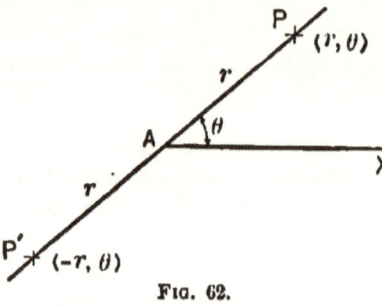

Fig. 62.

The angle which a line AP makes with AX is denoted by θ. θ is positive when the angle is conceived to be generated by a line starting from coincidence with AX turning about A anti-clockwise; θ is negative when generated by a line turning about A clockwise. When θ is given, a line through A is determined. On this line a point is determined by giving the distance and direction of the point from A. The direction from A is indicated by calling distances measured from A in the direction AP of the side of the angle θ positive, those measured in the opposite direction negative. The point P is denoted by the symbol (r, θ), the point P' by the symbol $(-r, \theta)$. The symbols $(r, \theta + 2\pi n)$, $(-r, \theta + (2n+1)\pi)$, where n is any integer, denote the same point. To every symbol (r, θ) there corresponds one point of the plane; to every point of the plane there corresponds an infinite number of symbols (r, θ). Under the restriction that r and θ are positive, and that the values of θ can differ only by less than 2π, there exists a one-to-one correspondence between the symbol (r, θ)

POLAR COORDINATES

and the points of the plane, the pole only excepted. r and θ are called the polar coordinates of the point.

1. Locate the points whose polar coordinates are $(2, 0)$; $(-3, 0)$; $(3, \tfrac{1}{2}\pi)$; $(-2, \pi)$; $(4, \tfrac{3}{2}\pi)$; $(-4, \tfrac{1}{2}\pi)$; $(4, \tfrac{7}{2}\pi)$; $(1, 1)$; $(-2, 1)$; $(-1, 0)$; $(1, 180°)$; $(-4, 45°)$; $(4, 225°)$; $(-4, 405°)$; $(0, 0)$; $(0, 45°)$; $(0, 225°)$.

2. Show that $r'^2 + r''^2 - 2\,r'r''\cos(\theta' - \theta'')$ is the distance between (r', θ'), (r'', θ'').

3. Find the distances between the following pairs of points, $(4, \tfrac{1}{2}\pi)$, $(3, \pi)$; $(8, \tfrac{1}{4}\pi)$, $(6, \tfrac{3}{4}\pi)$; $(2\sqrt{2}, -\tfrac{1}{4}\pi)$, $(1, \tfrac{1}{4}\pi)$; $(0, 0)$, $(10, 45°)$; $(5, 45°)$, $(10, 90°)$; $(-6, 120°)$, $(-8, 30°)$.

Art. 37. — Polar Equations of Geometric Figures

The conditions to be satisfied by a moving point can sometimes be more readily expressed in polar coordinates than in rectilinear coordinates. If a point moves in the XY-plane in such a manner that its distance from the origin varies directly as the angle included by the X-axis and the line from the origin to the moving point, the rectangular equation of the locus is $\sqrt{x^2 + y^2} = a \tan^{-1}\tfrac{y}{x}$, the polar equation $r = a\theta$.

Desired information about a curve is often obtained more directly from the polar equation than from the rectilinear equation of the curve. This is especially the case when the distances from a fixed point to various points of the curve are required. Thus if the orbit of a comet is a parabola with the sun at the focus, the comet's distance from the sun at any time is obtained directly from the polar equation of the parabola.

Art. 38. — Polar Equation of Straight Line

A straight line is determined when the length of the perpendicular from the pole to the line and the angle included by this perpendicular and the polar axis are given. Call the perpendicular p, the angle α, and let (r, θ) be any point of the line. The equation

$$r = \frac{p}{\cos(\theta - \alpha)}$$

expresses a relation satisfied by the coordinates r, θ of every point of the straight line and by the coordinates of no other point; that is, this is the equation of the straight line. For $\theta = 0$,

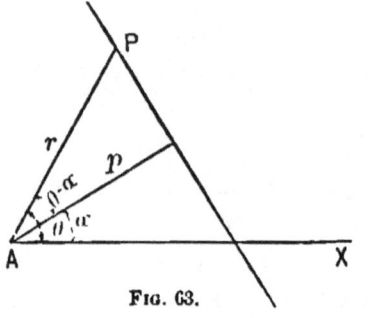

Fig. 63.

$r = \frac{p}{\cos \alpha}$; for $\theta = \alpha$, $r = p$; from $\theta = 0$ to $\theta = 90° + \alpha$, r is positive; from $\theta = 90° + \alpha$ to $\theta = 270° + \alpha$, r is negative; from $\theta = 270° + \alpha$ to $\theta = 360°$, r is again positive. For $\theta = 90° + \alpha$ and for $\theta = 270° + \alpha$, $r = \pm \infty$. These results obtained from the equation agree with facts observed from the figure.

A straight line is also determined by its intercept on the polar axis and the angle the line makes with the polar axis.

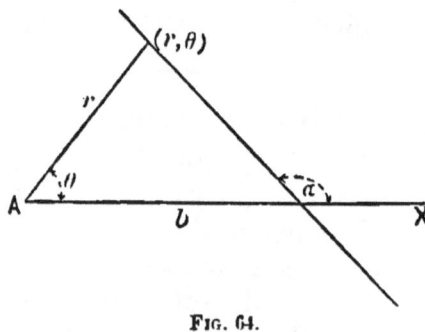

Fig. 64.

Call the intercept b, the angle α, and let (r, θ) be any point in the line. Then $r = \frac{b \sin \alpha}{\sin(\alpha - \theta)}$ is the equation of the line. For $\theta = 0$, $r = b$; r is positive from $\theta = 0$ to $\theta = \alpha$; negative from $\theta = \alpha$ to $\theta = 180° + \alpha$; again positive from $\theta = 180° + \alpha$ to $\theta = 360°$. For $\theta = \alpha$ and $\theta = 180° + \alpha$, $r = \pm \infty$. These results may be obtained from the equation or from the figure.

Art. 39. — Polar Equation of Circle

The equation of a circle whose radius is R when the pole is at the center, the polar axis a diameter, is $r = R$.

When the pole is on the circumference, the polar axis a diameter, the equation of the circle is $r = 2R\cos\theta$.

r is positive from $\theta = 0°$ to $\theta = 90°$, negative from $\theta = 90°$ to $\theta = 270°$, and again positive from $\theta = 270°$ to $\theta = 360°$. The entire circumference is traced from $\theta = 0°$ to $\theta = 180°$, and traced a second time from $\theta = 180°$ to $\theta = 360°$.

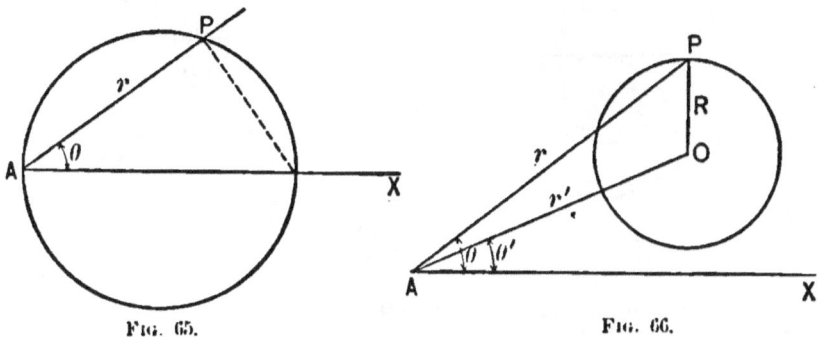

Fig. 65.   Fig. 66.

The polar equation of a circle radius R, center (r', θ'), current coordinates r, θ, is $r^2 - 2r'r\cos(\theta - \theta') = R^2 - r'^2$, whence $r = r'\cos(\theta - \theta') \pm \sqrt{R^2 - r'^2\sin^2(\theta - \theta')}$. r is real and has two unequal values when $\sin^2(\theta - \theta') < \dfrac{R^2}{r'^2}$; that is, when $-\dfrac{R}{r'} < \sin(\theta - \theta') < \dfrac{R}{r'}$; these values of r become equal, and the radius vector tangent to the circle, when $\sin(\theta - \theta') = \pm\dfrac{R}{r'}$; r is imaginary when $\sin^2(\theta - \theta') > \dfrac{R^2}{r_1^2}$.

Art. 40. — Polar Equations of the Conic Sections

Take the focus as pole, the axis of the conic section as polar axis. From the definition of a conic section

$$r = e \cdot DE = e(DA + AE) = e\left(\frac{p}{e} + r\cos\theta\right).$$

Hence $\quad r = p + er\cos\theta, \quad r = \dfrac{p}{1 - e\cos\theta}.$

Since in the parabola $e = 1$, the polar equation of the parabola is $r = \dfrac{p}{1 - \cos \theta}$.

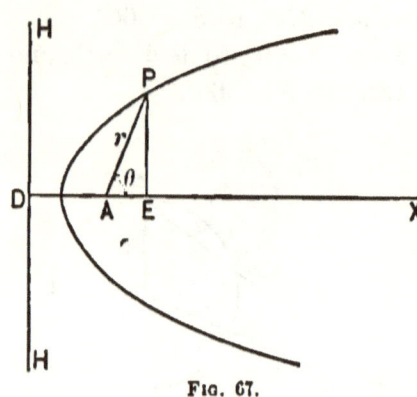

Fig. 67.

In the ellipse and hyperbola the numerical value of the semi-parameter p is

$$a(1 - e^2);$$

hence the polar equation of ellipse and hyperbola is

$$r = \dfrac{a(1 - e^2)}{1 - e \cos \theta}.$$

In the ellipse e is less than unity, and r is therefore always positive. For $\theta = 0$, $r = a(1 + e)$, showing that the pole is at the left-hand focus.

In the hyperbola e is greater than unity, and r is positive from $\theta = 0$ to $\theta = \cos^{-1}\dfrac{1}{e}$ in the first quadrant, negative from $\theta = \cos^{-1}\dfrac{1}{e}$ in the first quadrant to $\theta = \cos^{-1}\dfrac{1}{e}$ in the fourth quadrant, again positive from $\theta = \cos^{-1}\dfrac{1}{e}$ in the fourth quadrant to $\theta = 360°$. r becomes infinite for $\theta = \cos^{-1}\dfrac{1}{e}$; hence lines through the focus making angles whose cosine is $\dfrac{1}{e}$ with the axis of the hyperbola are parallel to the asymptotes of the hyperbola.

Problems. — 1. The length of the perpendicular from the pole to a straight line is 5; this perpendicular makes with the polar axis an angle of 45°. Find the equation of the line and discuss it.

2. Derive and discuss the polar equation of the straight line parallel to the polar axis and 8 above it.

3. Derive and discuss the equation of the straight line at right angles to the polar axis, and intersecting the polar axis 4 to the right of the pole.

4. Derive and discuss the equation of the circle, radius 5, center $(10, \tfrac{1}{2}\pi)$.

POLAR COORDINATES 75

5. Derive and discuss the equation of the circle, radius 10, center $(5, \frac{3}{2}\pi)$.

6. Derive and discuss the equation of the circle, radius 8, center $(10, \frac{1}{4}\pi)$.

7. Derive and discuss the equation of the circle, radius 10, center $(15, \pi)$.

8. Derive and discuss the equation of the circle, radius 10, center $(10, \frac{1}{2}\pi)$.

9. Find the polar equation of the parabola whose parameter is 12.

10. Find the polar equation of the ellipse whose axes are 8 and 6.

11. Find the polar equation of the ellipse, parameter 10, eccentricity $\frac{1}{2}$.

12. Find the polar equation of the ellipse, transverse axis 10, eccentricity $\frac{3}{5}$.

13. Find the polar equation of the hyperbola whose axes are 8 and 6.

14. Find polar equation of hyperbola, transverse axis 12, parameter 6.

15. Find polar equation of hyperbola, transverse axis 8, distance between foci 10.

16. Find the equation of the locus of a point moving in such a manner that the product of the distances of the point from two fixed points is always the square of the half distance between the fixed points. This curve is called the lemniscate of Bernoulli.

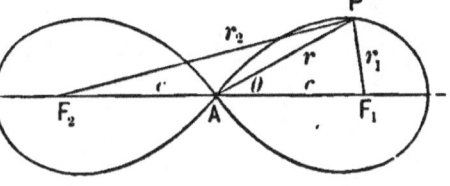

Fig. 68.

By definition $r_1 r_2 = c^2$. From the figure $r_1^2 = r^2 + c^2 - 2cr \cos\theta$, $r_2^2 = r^2 + c^2 + 2cr \cos\theta$, hence $r_1^2 r_2^2 = r^4 + 2c^2 r^2 + c^4 - 4c^2 r^2 \cos^2\theta = c^4$ and $r^2 = 2c^2(2\cos^2\theta - 1)$, $r^2 = 2c^2 \cos(2\theta)$.

Corresponding pairs of values of r_1 and r_2 may be found by drawing a circle with radius c, to this circle a tangent whose length is c. The distances from the end of the tangent to the points of intersections of the straight lines through the end of the tangent with the circumference are corresponding values of r_1 and r_2, for $TS \cdot TR = c^2$. The intersections of arcs struck off from F_1 and F_2 as centers with radii TS and TR determine points of the lemniscate.

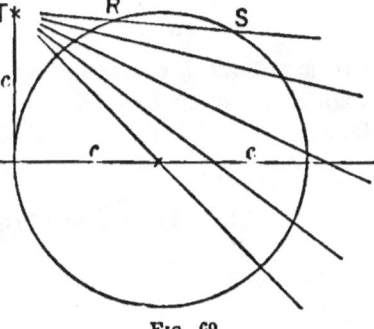

Fig. 69.

17. A bar turns around and slides on a fixed pin in such a manner that a constant length projects beyond a fixed straight line. Find the equation of the curve traced by the end of the bar. This curve is called the conchoid of Nicomedes.

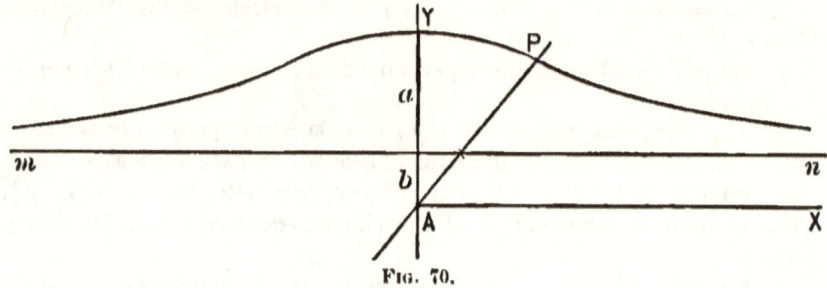

Fig. 70.

Take the fixed point A as focus, the line AX, parallel to the fixed line mn, as polar axis. Call the distance from the pole to the fixed line b, the constant length projecting beyond the fixed line a. Then

$$r = \frac{b}{\sin \theta} + a.$$

The conchoid is used to trisect an angle graphically. Let GAH be the angle. From any point B in one side of the angle draw a perpendicular mn to the other. With vertex of angle as fixed point, mn as fixed line, and $FG = 2\,BA$ as constant distance, construct a conchoid. At B erect perpendicular BC to mn, and join its point of intersection with conchoid C and A by a straight line.

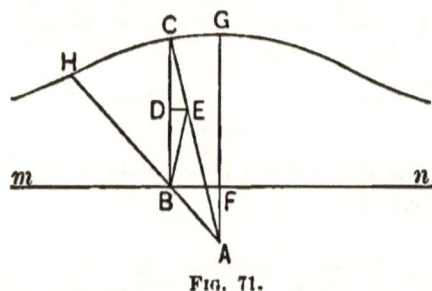

Fig. 71.

GAC is $\tfrac{1}{3} GAH$, for, drawing through D, the middle point of BC, a parallel to mn and joining B and E, the triangles ABE and BEC are isosceles. Hence $BAC = BEA = 2\,BCA = 2\,GAC$.

Art. 41.—Plotting of Polar Equations

Example.—Plot $r = 10 \cos \theta$.

$\theta =$	0	$\tfrac{1}{4}\pi$	$\tfrac{1}{2}\pi$	$\tfrac{3}{4}\pi$	π	$\tfrac{5}{4}\pi$	$\tfrac{3}{2}\pi$	$\tfrac{7}{4}\pi$	2π
$r =$	10	$5\sqrt{2}$	0	$-5\sqrt{2}$	-10	$-5\sqrt{2}$	0	$5\sqrt{2}$	10

If the number of points located from $\theta = 0$ to $\theta = 2\pi$ is indefinitely increased, the polygon formed by joining the successive points approaches the circumference of a circle as its limit. The form of the equation shows at once that the locus is a circle whose radius is 5.

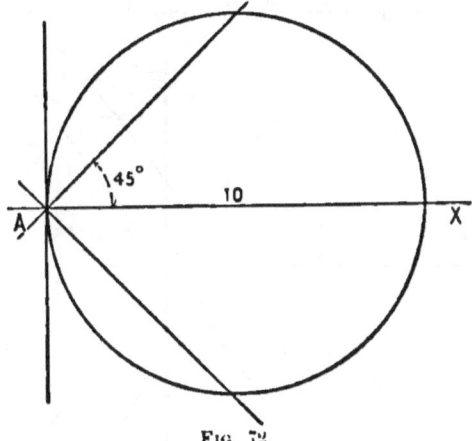

Fig. 72.

EXAMPLE. — Plot $r = a\theta$.

$\theta =$	-4	-3	-2	-1	0	1	2	3	4	$\cdots$
$r =$	$-4a$	$-3a$	$-2a$	$-a$	0	a	$2a$	$3a$	$4a$	$\cdots$

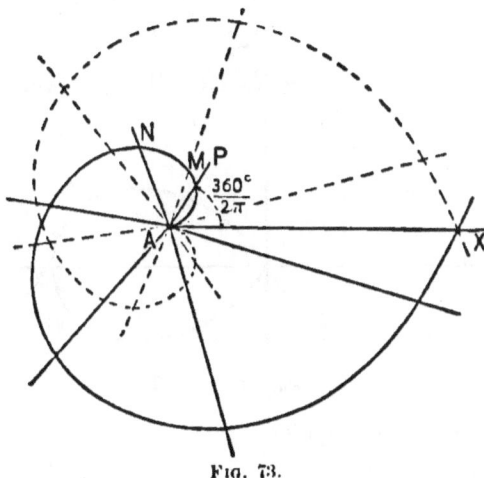

The curve is called the spiral of Archimedes. In rectangular coordinates the equation of this spiral is transcendental.

Fig. 73.

EXAMPLE. — Plot $r = \dfrac{4}{3 - 5\cos\theta}$.

$\theta = 0$	$\cos^{-1}\tfrac{3}{5}$	$\tfrac{1}{2}\pi$	π	$\tfrac{3}{2}\pi$	$\cos^{-1}\tfrac{3}{5}$	2π
$r = -2$	$\mp\infty$	$\tfrac{4}{3}$	$\tfrac{1}{2}$	$\tfrac{4}{3}$	$\pm\infty$	-2

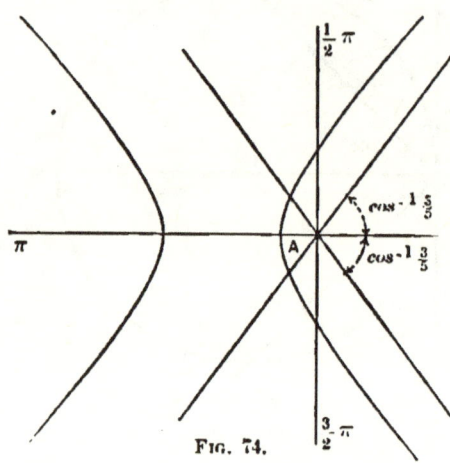

From $\theta = 0$ to $\theta = \cos^{-1}\frac{3}{5}$, r varies continuously from -2 to $-\infty$; from $\theta = \cos^{-1}\frac{3}{5}$ to $\theta = \pi$, r decreases continuously from $+\infty$ to $+\frac{1}{2}$; from $\theta = \pi$ to $\theta = \cos^{-1}\frac{3}{5}$ in the fourth quadrant, r increases continuously from $\frac{1}{2}$ to $+\infty$; from $\theta = \cos^{-1}\frac{3}{5}$ in the fourth quadrant to $\theta = 2\pi$, r increases from $-\infty$ to -2. r is discontinuous for $\theta = \cos^{-1}\frac{3}{5}$. This equation represents an hyperbola whose less focal distance is $\frac{1}{2}$, greater focal distance 2, semi-parameter $\frac{4}{3}$, eccentricity $\frac{5}{3}$.

Fig. 74.

EXAMPLE. — Plot $r^2 = 8 \cos(2\theta)$.

$\theta =$	$0°$	$22\frac{1}{2}°$	$45°$	....	$135°$	$157\frac{1}{2}°$	$180°$
$r =$	± 2.828	± 2.378	0	imaginary	0	± 2.378	± 2.828

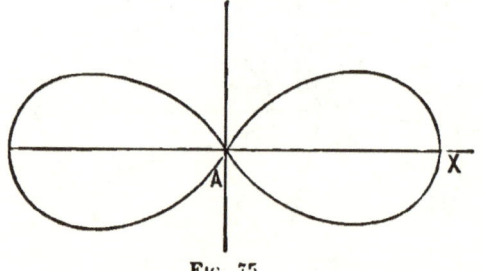

From $180°$ to $360°$ the curve is traced a second time. The pole is a center of symmetry of the curve.

Fig. 75.

Problems. — Plot

1. $r = \dfrac{3}{\cos\theta}$.
2. $r = \dfrac{-4}{\cos(\theta - \frac{1}{4}\pi)}$.
3. $r = a\cos(3\theta)$.
4. $r = 2\cos\theta$.
5. $r = a\sin(2\theta)$.
6. $r = a\cos(3\theta)$.
7. $r = a\sin(3\theta)$.
8. $r = a\sin(4\theta)$.
9. $r = a\sin(5\theta)$.
10. $r = \dfrac{a}{\theta}$, the reciprocal spiral.
11. $r = a^\theta$, the logarithmic spiral.

12. $r = \dfrac{a}{\theta^{\frac{1}{2}}}$, the lituus.

13. $r = \dfrac{2}{1 - \cos \theta}$.

14. $r = \dfrac{5}{2 - 3 \cos \theta}$.

15. $r = \dfrac{4}{3 - 2 \cos \theta}$.

16. $r = \dfrac{10}{1 + \cos \theta}$.

17. $r = a(1 + \cos \theta)$, the cardioid.
18. $r = 4(1 - \cos \theta)$.
19. $r = 5 + 2 \sin \theta$.

20. $r = 2p \cot \theta \csc \theta$.

21. $r = \dfrac{4 \cos \theta}{1 + 3 \sin^2 \theta}$.

22. $r = \dfrac{4 \cos \theta}{1 - 5 \sin^2 \theta}$.

23. $r = \dfrac{3 \sin \theta \cos \theta}{\sin^3 \theta + \cos^3 \theta}$.

24. $r = a(\sin 2\theta + \cos 2\theta)$.
25. $r^2 \cos(2\theta) = 4$.
26. $r^2 \sin(2\theta) = 8$.
27. $r^{\frac{1}{2}} \cos \tfrac{1}{2} \theta = 2$.
28. $r^2 = 16 \sin(2\theta)$.

Art. 42.—Transformation from Rectangular to Polar Coordinates

If the rectangular equation of a geometric figure is given, and the polar equation is desired, find the values of the rectangular coordinates x and y of any point in terms of the polar coordinates r and θ of the same point; substitute in the rectangular equation $f(x, y) = 0$, and the resulting equation $F(r, \theta) = 0$ is the polar equation of the figure.

Let the pole A' referred to the rectangular coordinates be (m, n), θ' the angle made by the polar axis with the X-axis. Then

$x = AD = m + r \cos(\theta + \theta')$,
$y = PD = n + r \sin(\theta + \theta')$.

When the pole is at the origin, and the polar axis coincides with the X-axis, these formulas become $x = r \cos \theta$, $y = r \sin \theta$.

If the polar equation of a geometric figure is given and

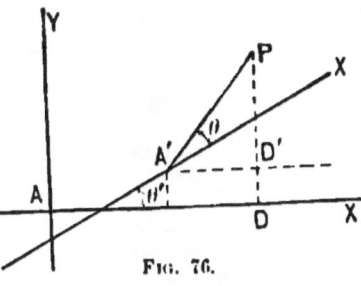

Fig. 76.

the rectangular equation is desired, find the values of the polar coordinates r and θ of any point in terms of the rectangular coordinates x and y of the same point; substitute in the polar equation $F(r, \theta) = 0$, and the resulting equation $f(x, y) = 0$ is the rectangular equation of the curve.

From the figure

$$r = \sqrt{(x - m)^2 + (y - n)^2}, \quad \cos(\theta + \theta') = \frac{x - m}{\sqrt{(x - m)^2 + (y - n)^2}},$$

$$\sin(\theta + \theta') = \frac{y - n}{\sqrt{(x - m)^2 + (y - n)^2}}.$$

When the origin is at the pole, and the X-axis coincides with the polar axis, these formulas become

$$r = \sqrt{x^2 + y^2}, \quad \cos\theta = \frac{x}{\sqrt{x^2 + y^2}}, \quad \sin\theta = \frac{y}{\sqrt{x^2 + y^2}}.$$

Problems. — Transform from rectangular coordinates to polar, pole at origin, polar axis coinciding with X-axis, and plot the locus from both equations.

1. $x^2 + y^2 = 25$.
2. $x^2 + y^2 - 10x = 0$.
3. $y^2 = 2px$.
4. $x^2 - y^2 = 25$.
5. $xy = 9$.
6. $y^2 = \frac{1}{4}(4x - x^2)$.
7. $y^2 = -\frac{1}{4}(4x - x^2)$.
8. $y^3 - 3xy + x^3 = 0$.
9. $(x^2 + y^2)^2 = a^2(x^2 - y^2)$.

Transform from polar coordinates to rectangular coordinates, X-axis coinciding with polar axis.

10. $r = a$, origin at pole.
11. $r = 10 \cos\theta$, origin at pole.
12. $r^2 = a^2 \cos(2\theta)$, origin at pole.
13. $r^{\frac{1}{2}} \cos\frac{1}{2}\theta = 2$, origin at pole.
14. $r = \frac{p}{1 - \cos\theta}$, pole at $(\frac{1}{2}p, 0)$.
15. $r = \frac{9}{4 - 5\cos\theta}$, pole at $(5, 0)$.
16. $r = \frac{9}{5 - 4\cos\theta}$, pole at $(4, 0)$.
17. $r^2 = \frac{\cos(2\theta)}{\cos^4\theta}$, origin at pole.
18. $r^2 \cos^4\theta = 1$, origin at pole.

CHAPTER VII

PROPERTIES OF THE STRAIGHT LINE

Art. 43. — Equations of the Straight Line

The various conditions determining a straight line give rise to different forms of the equation of a straight line.

I. The equation of the straight line determined by the two points (x', y'), (x'', y'').

The similarity of the triangles $PP'D$ and $P'P'''D'$ is the geometric condition which locates the point $P(x, y)$ on the straight line through $P'(x', y')$ and $P''(x'', y'')$. This condition leads to the equation $y - y' = \frac{y' - y''}{x' - x''}(x - x')$. In rec-

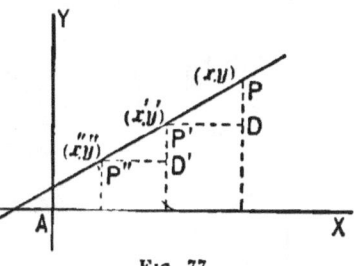

Fig. 77.

tangular coordinates $\frac{y' - y''}{x' - x''} = \tan \alpha$, where α is the angle the line makes with the X-axis. In oblique coordinates $\frac{y' - y''}{x' - x''} = \frac{\sin \alpha}{\sin(\beta - \alpha)}$, where β is the angle between the axes, α the angle the line makes with the X-axis.

II. The equation of a straight line through a given point (x', y') and making a given angle α with the X-axis is $y - y' = \tan \alpha (x - x')$. If the point (x', y') is the intersection $(0, n)$ of the line with the X-axis and $\tan \alpha = m$, the equation becomes $y = mx + n$, the slope equation of a straight line.

On the straight line $y - y' = \tan \alpha (x - x')$ the coordinates of the point whose distance from (x', y') is d, are $x = x' + d \cos \alpha$, $y = y' + d \sin \alpha$.

III. The equation of the straight line whose intercepts on the axes are a and b.

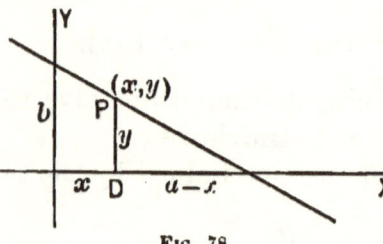

Fig. 78.

Let (x, y) be any point in the line. From the figure $\frac{a-x}{a} = \frac{y}{b}$, which reduces to $\frac{x}{a} + \frac{y}{b} = 1$, the intercept equation of a straight line.

IV. When the length p and the inclination α to the X-axis of the perpendicular from the origin to the straight line are given.

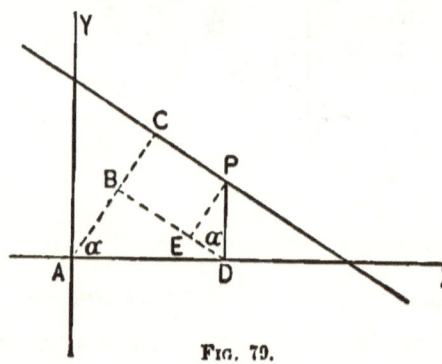

Fig. 79.

Let (x, y) be any point in the straight line. From the figure, $AB + BC = p$, hence
$$x \cos \alpha + y \sin \alpha = p.$$
This is the normal equation of a straight line.

The different forms of the equation of a straight line can be obtained from the general first degree equation in two variables $Ax + By + C = 0$, which always represents a straight line.

(a) Suppose the two points (x', y'), (x'', y'') to lie in the line represented by the equation $Ax + By + C = 0$. The elimination of A, B, C from (1) $Ax + By + C = 0$, (2) $Ax' + By' + C = 0$, (3) $Ax'' + By'' + C = 0$ by subtracting (2) from (1) and (3) from (1), and dividing the resulting equations gives
$$y - y' = \frac{y' - y''}{x' - x''}(x - x').$$

(b) Calling the intercept of the line $Ax + By + C = 0$ on the X-axis a, on the Y-axis b, for $y = 0$, $x = -\frac{C}{A} = a$, for $x = 0$, $y = -\frac{C}{B} = b$. Substituting in the equation $Ax + By + C = 0$, there results $\frac{x}{a} + \frac{y}{b} = 1$.

PROPERTIES OF THE STRAIGHT LINE

(c) The equation $Ax + By + C = 0$ may be written

$$y = -\frac{A}{B}x - \frac{C}{B},$$

which is of the form $y = mx + n$.

(d) Let $Ax + By + C = 0$ and $x \cos \alpha + y \sin \alpha = p$ represent the same line. There must exist a constant factor m such that $mAx + mBy + mC = 0$ and $x \cos \alpha + y \sin \alpha - p = 0$ are identical. From this identity $mA = \cos \alpha$, $mB = \sin \alpha$, $mC = -p$. The first two equations give $m^2A^2 + m^2B^2 = 1$, hence $m = \dfrac{1}{\sqrt{A^2 + B^2}}$. That is,

$$\frac{A}{\sqrt{A^2 + B^2}}x + \frac{B}{\sqrt{A^2 + B^2}}y + \frac{C}{\sqrt{A^2 + B^2}} = 0$$

is the normal form of the equation of the straight line represented by $Ax + By + C = 0$.

The nature of the problem generally indicates what form of the equation of the straight line it is expedient to use.

Problems. — 1. Write the equation of the straight line through the points $(2, 3)$, $(-1, 4)$.

2. Write the equation of the straight line through $(-2, 3)$, $(0, 4)$.

3. Write the intercept equation of the straight line through $(4, 0)$, $(0, 3)$.

4. Write the equation of the straight line whose perpendicular distance from the origin is 5, this perpendicular making an angle of 30° with the X-axis.

5. Write the equation $\dfrac{x}{2} + \dfrac{y}{3} = 1$ in the slope form.

6. Write the equation $2x - 3y = 5$ in the normal form.

7. Write the equation of the straight line through $(4, -3)$, making an angle of 135° with the X-axis.

8. On the straight line through $(-2, 3)$, making an angle of 30° with the X-axis, find the coordinates of the point whose distance from $(-2, 3)$ is 6.

9. The vertices of a triangle are $(3, 7)$, $(5, -1)$, $(-3, 5)$. Write equations of medians.

Art. 44. — Angle between Two Lines

Let V be the angle between the straight lines $y = mx + n$, $y = m'x + n'$. From the figure $V = \alpha - \alpha'$, hence

$$\tan V = \frac{\tan \alpha - \tan \alpha'}{1 + \tan \alpha \tan \alpha'}.$$

Since $\tan \alpha = m$, $\tan \alpha' = m'$, $\tan V = \dfrac{m - m'}{1 + mm'}$. When the lines are parallel, $V = 0$, which

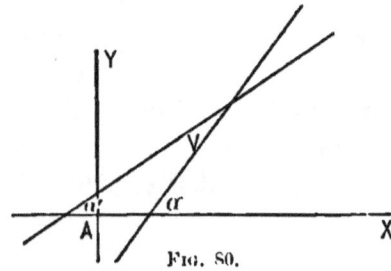

Fig. 80.

requires that $m = m'$. When the lines are perpendicular, $V = 90°$, which requires that $1 + mm' = 0$, or $m' = -\dfrac{1}{m}$.

If the equations of the lines are written in the form $Ax + By + C = 0$, $A'x + B'y + C' = 0$, $\tan V = \dfrac{A'B - AB'}{AA' + BB'}$. The lines are parallel when $A'B - AB' = 0$, perpendicular when $AA' + BB' = 0$.

The equation of the straight line through (x', y') perpendicular to $y = mx + n$ is $y - y' = -\dfrac{1}{m}(x - x')$.

The equation of the straight line through (x', y') parallel to $y = mx + n$ is $y - y' = m(x - x')$.

Let the straight line $y - y' = \tan \alpha'(x - x')$ through the point (x', y') make an angle θ with the line $y = mx + n$. From the figure, $\alpha' = \theta + \alpha$. Hence

$$\tan \alpha' = \frac{\tan \theta + \tan \alpha}{1 - \tan \theta \tan \alpha} = \frac{\tan \theta + m}{1 - m \tan \theta},$$

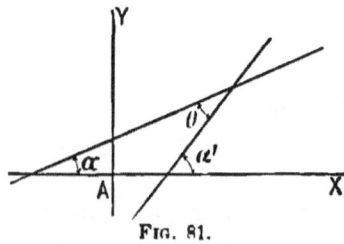

Fig. 81.

since $\tan \alpha = m$. Therefore the equation of the line through (x', y') making an angle θ with the line $y = mx + n$ is

$$y - y' = \frac{\tan \theta + m}{1 - m \tan \theta}(x - x').$$

PROPERTIES OF THE STRAIGHT LINE

Problems. — 1. Find the angle the line $\frac{x}{4} - \frac{y}{3} = 1$ makes with the X-axis.

2. Find the angle between the lines $2x + 3y = 7$, $\frac{1}{2}x + \frac{1}{3}y = 1$.

3. Find the equation of the line through $(4, -2)$ parallel to $5x - 7y = 10$.

4. Find the equation of the line through $(1, 3)$ parallel to the line through $(2, 1)$, $(-3, 2)$.

5. Find the equation of the line through the origin perpendicular to $3x - y = 5$.

6. Find the equation of the line through $(2, -3)$ perpendicular to $\frac{3}{4}x - \frac{1}{2}y = 1$.

7. Find the equation of the line through $(0, -5)$ perpendicular to the line through $(4, 5)$, $(2, 0)$.

8. The vertices of a triangle are $(4, 0)$, $(5, 7)$, $(-6, 3)$. Find the equations of the perpendiculars from the vertices to the opposite sides.

9. The vertices of a triangle are $(3, 5)$, $(7, 2)$, $(-5, -4)$. Find the equations of the perpendiculars to the sides at their middle points.

10. Write equation of line through $(2, 5)$, making angle of $45°$ with $2x - 3y = 6$.

ART. 45. — DISTANCE FROM A POINT TO A LINE

Write the equation of the given line in the normal form $x \cos \alpha + y \sin \alpha - p = 0$. Through the given point $P(x', y')$ draw a line parallel to the given line. The normal equation of this parallel line is

$$x \cos \alpha + y \sin \alpha = AP'.$$

Since (x', y') is in this line,

$$x' \cos \alpha + y' \sin \alpha = AP'.$$

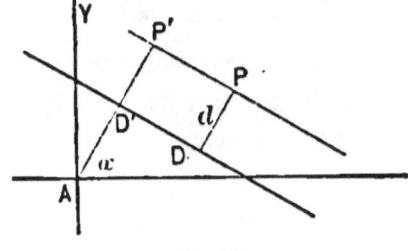

Fig. 82.

Subtracting $p = AD'$, there results $x' \cos \alpha + y' \sin \alpha - p = PD$; that is, the perpendicular distance from the point (x', y') to the line $x \cos \alpha + y \sin \alpha - p = 0$

is $x'\cos\alpha + y'\sin\alpha - p$. The manner of obtaining this result shows that the perpendicular PD is positive when the point P and the origin of coordinates lie on different sides of the given line; negative when the point P and the origin lie on the same side of the given line.

The perpendicular distance from (x', y') to $Ax + By + C = 0$ is found by writing this equation in the normal form

$$\frac{A}{\sqrt{A^2 + B^2}}x + \frac{B}{\sqrt{A^2 + B^2}}y + \frac{C}{\sqrt{A^2 + B^2}} = 0$$

and applying the former result to be $PD = \dfrac{Ax' + By' + C}{\sqrt{A^2 + B^2}}$.

This formula determines the length of the perpendicular; the algebraic sign to be prefixed, which indicates the relative positions of origin, point, and line, must be determined as before.

Problems. — 1. Find distance from $(-2, 3)$ to $3x + 5y = 15$.

2. Find distance from origin to $\frac{2}{3}x - \frac{3}{5}y = 7$.

3. Find distance from $(4, -5)$ to line through $(2, 1)$, $(-3, 5)$.

4. Find distance from $(3, 7)$ to $\dfrac{x - 3y}{4} = \dfrac{7y - 5x}{2}$.

5. The vertices of a triangle are $(3, 2)$, $(-4, 2)$, $(5, -7)$. Find lengths of perpendiculars from vertices to opposite sides.

6. The sides of a triangle are $y = 2x + 5$, $\dfrac{x}{3} - \dfrac{y}{2} = 1$, $4x - 7y = 12$. Find lengths of perpendiculars from vertices to opposite sides.

7. The sides of a triangle are $y = 2x + 3$, $y = -\frac{1}{2}x + 2$, $y = x - 5$. Find area of triangle.

Art. 46. — Equations of Bisectors of Angles

Let the sides of the angles be $Ax + By + C = 0$, $A'x + B'y + C' = 0$. The bisector ab is the locus of all points equidistant from the given lines such that the points and the origin lie either on the same side of each of the two given lines or on different sides of each of the two given lines. In either case the perpendiculars from any point (x, y) of the bisector to the given lines have the same sign, and the equation of the bisector is

$$\frac{Ax + By + C}{\sqrt{A^2 + B^2}} = \frac{A'x + B'y + C'}{\sqrt{A'^2 + B'^2}}.$$

The bisector cd is the locus of all points equidistant from the given lines and situated on the same side of one of the given lines with the origin, while the other line lies between the points of the bisector and the origin. The perpendiculars from any point (x, y) of the bisector cd to the given lines are therefore numerically equal but with opposite signs, and the equation of the bisector cd is
$$\frac{Ax + By + C}{\sqrt{A^2 + B^2}} = -\frac{A'x + B'y + C'}{\sqrt{A'^2 + B'^2}}.$$

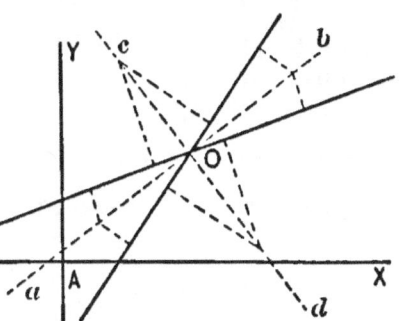

Fig. 83.

Problems. — **1.** Find the bisectors of the angles whose sides are $3x + 4y = 5$, $5x - 7y = 2$.

2. Find the bisectors of the angles whose sides are $\tfrac{1}{3}x - \tfrac{1}{2}y = 1$, $y = 2x - 3$.

3. Find locus of all points equidistant from the lines $2x + 7y = 10$, $8x - 5y = 15$.

4. The sides of a triangle are $5x + 3y = 9$, $\tfrac{1}{2}x + \tfrac{1}{3}y = 1$, $y = 3x - 10$. Find the bisectors of the angles.

5. The sides of a triangle are $7x + 5y = 14$, $10x - 15y = 21$, $y = 3x + 7$. Find the center of the inscribed circle.

Art. 47. — Lines through Intersection of Given Lines

Let (1) $Ax + By + C = 0$ and (2) $A'x + B'y + C'' = 0$ be the given lines. Then (3) $Ax + By + C + k(A'x + B'y + C') = 0$, where k is an arbitrary constant, represents a straight line through the point of intersection of (1) and (2). For equation (3) is of the first degree, hence it represents a straight line. Equation (3) is satisfied when (1) and (2) are satisfied simultaneously, hence the line represented by (3) contains the point of intersection of the lines represented by equations (1) and (2).

If the line $Ax + By + C + k(A'x + B'y + C') = 0$ is to contain the point (x', y'), k becomes $-\dfrac{Ax' + By' + C}{A'x' + B'y' + C'}$. Hence

$$Ax + By + C - \dfrac{Ax' + By' + C}{A'x' + B'y' + C'}(A'x + B'y + C') = 0$$

is the equation of the line through (x', y'), and the intersection of (1) and (2).

If the equations of the given lines are written in the normal form, (1) $x \cos \alpha + y \sin \alpha - p = 0$, (2) $x \cos \alpha' + y \sin \alpha' - p' = 0$, the k of the line through their point of intersection

(3) $x \cos \alpha + y \sin \alpha - p + k(x \cos \alpha' + y \sin \alpha' - p') = 0$

has a direct geometric interpretation. $k = -\dfrac{x \cos \alpha + y \sin \alpha - p}{x \cos \alpha' + y \sin \alpha' - p'}$, that is, k is the negative ratio of the distances from any point (x, y) of the line (3) to the lines (1) and (2).

Problems. — 1. Find the equation of the line through the origin and the point of intersection of $3x - 4y = 5$ and $2x + 5y = 8$.

2. Find the equation of the locus of the points whose distances from the lines $\frac{2}{3}x - 5y + 2 = 0$, $\dfrac{x}{3} - \dfrac{y}{5} = 1$ are in the ratio of 2 to 3.

3. Find the equation of the line through $(-2, 3)$ and the intersection of the lines $8x - 5y = 15$, $3x + 10y = 8$.

Art. 48. — Three Points in a Straight Line

Let the three points (x', y'), (x'', y''), (x''', y''') lie in a straight line. The equation of the straight line through the first two points is $y - y' = \dfrac{y' - y''}{x' - x''}(x - x')$. By hypothesis the point

(x''', y''') lies in this line, hence $y''' - y' = \dfrac{y' - y''}{x' - x''}(x''' - x')$. Simplifying, (1) $x'y''' - x''y''' + x''y' - x'''y' + x'y'' - x'''y'' = 0$. When this equation is satisfied the three points lie in a straight line, whether the coordinates are rectangular or oblique. Notice that (1) expresses the condition that the area of the triangle whose vertices are (x', y'), (x'', y''), (x''', y''') is zero.

Problems. — 1. In a parallelogram each of the two sides through a vertex is prolonged a distance equal to the length of the other side. Prove that the opposite vertex of the parallelogram and the ends of the produced sides lie in a straight line.

2. In a jointed parallelogram on two sides through a common vertex two points are taken in a straight line with the opposite vertex. Show that these three points are in a straight line however the parallelogram is distorted.

Art. 49. — Three Lines Through a Point

Let the three lines $Ax + By + C = 0$, $A'x + B'y + C' = 0$, $A''x + B''y + C'' = 0$ pass through a common point. Make the first two of these equations simultaneous, solve for x and y, and substitute the values found in the third equation. There results

$$AB'C'' + A'B''C + A''BC' - A''B'C - A'BC'' - AB''C' = 0,$$

which is the condition necessary for the intersection of the given lines.

The three lines necessarily have a common point if constants κ_1, κ_2, κ_3 can be found such that $\kappa_1(Ax + By + C) + \kappa_2(A'x + B'y + C') + \kappa_3(A''x + B''y + C'') = 0$ is identically satisfied. For the values of x and y which satisfy $Ax + By + C = 0$, and $A'x + B'y + C' = 0$ simultaneously must then also satisfy $A''x + B''y + C'' = 0$; that is, the point of intersection of the first two lines lies in the third line.

The second criterion is frequently more convenient of application than the first.

Problems.—1. The bisectors of the angles of a triangle pass through a common point.

Let the normal equations of the three sides of the triangle be
$$x\cos\alpha + y\sin\alpha - p_1 = 0, \quad x\cos\beta + y\sin\beta - p_2 = 0, \quad x\cos\gamma + y\sin\gamma - p_3 = 0.$$

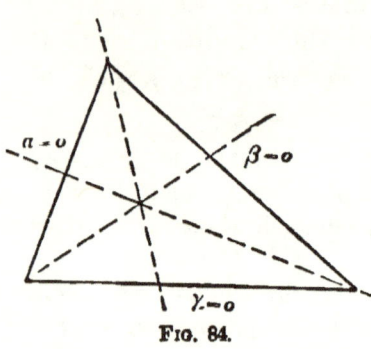

Fig. 84.

Denote the left-hand members of these equations by α, β, γ. Then $\alpha = 0$, $\beta = 0$, $\gamma = 0$ represent the sides of the triangle, and α, β, γ evaluated for the coordinates of any point (x, y) are the distances from this point to the sides of the triangle. Hence the equations of the bisectors of the angles are $\alpha - \beta = 0$, $\beta - \gamma = 0$, $\gamma - \alpha = 0$. The sum of the equations of the bisectors is identically zero, therefore the bisectors pass through a common point.

2. The medians of a triangle pass through a common point.

For every point in the median through C, $\dfrac{\alpha}{\sin B} = \dfrac{\beta}{\sin A}$, hence $\alpha \sin A - \beta \sin B = 0$ is the equation of the median through C. Similarly the equation of the median through B is found to be
$$\gamma \sin C - \alpha \sin A = 0;$$
of the median through A, $\beta \sin B - \gamma \sin C = 0$. The sum of these equations vanishes identically.

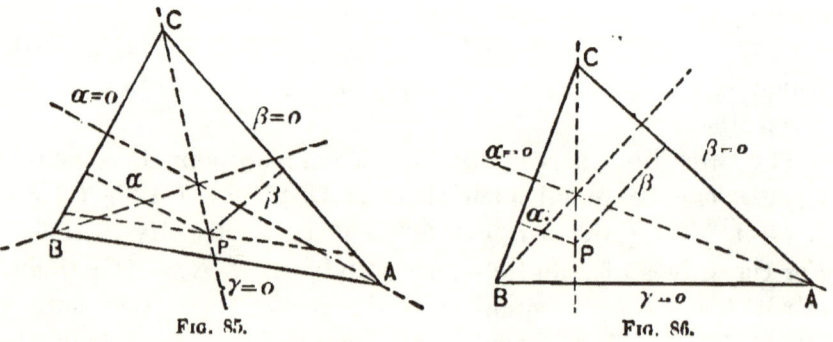

Fig. 85.   Fig. 86.

3. The perpendiculars from the vertices of a triangle to the opposite sides pass through a common point.

The equation of the perpendicular through C is $\alpha \cos A - \beta \cos B = 0$; through B, $\gamma \cos C - \alpha \cos A = 0$; through A, $\beta \cos B - \gamma \cos C = 0$.

Art. 50. — Tangent to Curve of Second Order

The general equation of the curve of the second order is $ax^2 + 2bxy + cy^2 + 2dx + 2ey + f = 0$. Let (x_0, y_0) be any point in the curve. The equation $y - y_0 = \tan \alpha (x - x_0)$ represents any line through (x_0, y_0). The line cuts the curve of the second order in two points and is a tangent when the two points coincide. The coordinates of any point in the straight line are $x = x_0 + l \cos \alpha$, $y = y_0 + l \sin \alpha$. The points of intersection of straight line and curve of second order are the points corresponding to the values of l satisfying the equation

$$(ax_0^2 + 2bx_0y_0 + cy_0^2 + 2dx_0 + 2ey_0 + f)$$
$$+ (2ax_0 \cos \alpha + 2bx_0 \sin \alpha + 2by_0 \cos \alpha$$
$$+ 2cy_0 \sin \alpha + 2d \cos \alpha + 2e \sin \alpha)l$$
$$+ (a \cos^2 \alpha + 2b \cos \alpha \sin \alpha + c \sin^2 \alpha)l^2 = 0.$$

Since (x_0, y_0) is in the curve, the absolute term of the equation vanishes. If the coefficient of the first power of l also vanishes, the equation has two roots equal to zero, that is the two points of intersection of $y - y_0 = \tan \alpha(x - x_0)$ with the curve coincide at (x_0, y_0) when

$$ax_0 \cos \alpha + bx_0 \sin \alpha + by_0 \cos \alpha + cy_0 \sin \alpha + d \cos \alpha + e \sin \alpha = 0.$$

The equation of the tangent is found by eliminating $\cos \alpha$ and $\sin \alpha$ from the three equations $x = x_0 + l \cos \alpha$, $y = y_0 + l \sin \alpha$, $ax_0 \cos \alpha + bx_0 \sin \alpha + by_0 \cos \alpha + cy_0 \sin \alpha + d \cos \alpha + e \sin \alpha = 0$. This elimination is best effected by multiplying the third equation by l, then substituting from the first two equations, $l \cos \alpha = x - x_0$, $l \sin \alpha = y - y_0$. The resulting equation reduces to

$$axx_0 + b(xy_0 + x_0y) + cyy_0 + d(x + x_0) + e(y + y_0) + f = 0.$$

The law of formation of the equation of the tangent from the equation of the curve is manifest.

Problems. — 1. Write the equation of the tangent to $x^2 + y^2 = r^2$ at (x_0, y_0).

2. Write the equation of the tangent to $\frac{x^2}{a^2} + \frac{y^2}{b^2} = 1$ at (x_0, y_0).

3. Write the equation of the tangent to $\frac{x^2}{a^2} - \frac{y^2}{b^2} = 1$ at (x_0, y_0).

4. Write the equation of the tangent to $y^2 = 2px$ at (x_0, y_0).

5. Find the equation of the tangents to $4x^2 + 9y^2 = 36$ at the points where $x = 1$.

6. At what point of $x^2 - y^2 = 1$ must a tangent be drawn to make an angle of $45°$ with the X-axis?

7. Find the angle under which the line $y = \frac{1}{2}x - 5$ cuts the circle $x^2 + y^2 = 49$.

8. Find the angle between the curves $y^2 = 6x$, $9y^2 + 4x^2 = 36$.

9. Find the equations of the normals to the ellipse, hyperbola, and parabola at the point (x_0, y_0) of the curve.

The normal to a curve at any point is the perpendicular through the point to the tangent to the curve at the point.

10. Where must the normal to $\frac{x^2}{9} + \frac{y^2}{4} = 1$ be drawn to make an angle of $135°$ with the X-axis?

11. Find the equation of the normal to $y^2 = 10x$ at $(10, 10)$.

12. Find equations of focal tangents to ellipse $\frac{x^2}{a^2} + \frac{y^2}{b^2} = 1$.

CHAPTER VIII

PROPERTIES OF THE CIRCLE

Art. 51. — Equation of the Circle

The equation of the circle referred to rectangular axes, radius R, center (a, b), is $(x - a)^2 + (y - b)^2 = R^2$. This equation represents all circles in the XY-plane. The equation expanded becomes $x^2 + y^2 - 2ax - 2by + a^2 + b^2 - R^2 = 0$, an equation of the second degree lacking the term in xy, and having the coefficients of x^2 and y^2 equal.

Conversely, every second degree equation lacking the term in xy, and having the coefficients of x^2 and y^2 equal, represents a circle when interpreted in rectangular coordinates. Such an equation has the form $x^2 + y^2 - 2ax - 2by + c = 0$, which when written in the form $(x - a)^2 + (y - b)^2 = a^2 + b^2 - c$, is seen to represent a circle of radius $(a^2 + b^2 - c)^{\frac{1}{2}}$, with center at (a, b). a, b, c are called the parameters of the circle, and the circle is spoken of as the circle (a, b, c).

When the center is at the origin, $a = 0$, $b = 0$, and the equation of the circle becomes $x^2 + y^2 = R^2$.

When the X-axis is a diameter, the Y-axis a tangent at the end of this diameter, the circle lying on the positive side of the Y-axis, $a = R$, $b = 0$, and the equation of the circle becomes $y^2 = 2Rx - x^2$.

Problems. — Write the equations of the following circles:

1. Center $(-2, 1)$, radius 5.
2. Center $(-5, 5)$, radius 5.
3. Center $(-10, 15)$, radius 5.
4. Center $(0, 0)$, radius 5.

5. Find equation of circle through $(0, 0)$, $(4, 0)$, $(0, 4)$.

6. Find center and radius of circle through $(2, -1)$, $(-2, 1)$, $(4, 5)$.

7. Find center and radius of circle $x^2 + y^2 + 4x - 10y = 7$.

8. Find center and radius of circle $x^2 + y^2 + 10x = 11$.

9. Does the line $3x - 5y = 12$ intersect the circle
$$x^2 + y^2 - 8x + 10y = 50?$$

10. Find the points of intersection of the circles
$$x^2 + y^2 - 10x + 6y = 20, \quad x^2 + y^2 + 4x - 15y = 25.$$

Art. 52. — Common Chord of Two Circles

The coordinates of the points of intersection of the circles
$$x^2 + y^2 - 2ax - 2by + c = 0, \quad x^2 + y^2 - 2a'x - 2b'y + c' = 0$$
satisfy the equation
$$(x^2 + y^2 - 2ax - 2by + c) - (x^2 + y^2 - 2a'x - 2b'y + c') = 0,$$
which reduces to
$$(a - a')x + (b - b')y + (c' - c) = 0.$$
This is the equation of the straight line through the points of intersection of the circles, that is the equation of the common chord of the circles.

The intersections of two circles may be a pair of real points, distinct or coincident, or a pair of conjugate imaginary points. Since the equation of the straight line through the points of intersection is in all cases real, it follows that the straight line through a pair of conjugate imaginary points is real.

Problems. — Write the equations of the common chords of the pairs of circles:

1. $x^2 + y^2 - 6x + 4y = 12, \quad x^2 + y^2 - 4x + 6y = 12$.
2. $x^2 + y^2 - 10x - 6y = 15, \quad x^2 + y^2 + 10x + 6y = 15$.
3. $x^2 + y^2 + 7x + 8y = 20, \quad x^2 + y^2 + 4x - 10y = 18$.

Art. 53.—Power of a Point

Let (x', y') be any point in the plane of the circle
$$(x - a)^2 + (y - b)^2 = R^2.$$
The equation of any straight line through (x', y') is
$$y - y' = \tan \alpha (x - x'),$$
and on this line the point at a distance d from (x', y') has for coordinates $x = x' + d \cos \alpha$, $y = y' + d \sin \alpha$. The distances from (x', y') to the points of intersection of line and circle are the values of d found by solving the equation
$$[(x' - a)^2 + (y' - b)^2 - R^2] + [2(x' - a)\cos \alpha$$
$$+ 2(y' - b)\sin \alpha] d + d^2 = 0.$$

Since the product of the roots of an equation equals numerically the absolute term of the equation, it follows that the product of the distances from the point (x', y') to the points of intersection of $y - y' = \tan \alpha (x - x')$ with the circle
$$(x - a)^2 + (y - b)^2 = R^2 \text{ is } (x' - a)^2 + (y' - b)^2 - R^2.$$

This product is independent of α; that is, it is the same for all lines through (x', y'). This constant product is called the power of the point (x', y') with respect to the circle.

The expression $(x' - a)^2 + (y' - b)^2 - R^2$ is the square of the distance from (x', y') to the center (a, b) minus the square of the radius. This difference, when the point (x', y') is without the circle, is the square of the tangent from the point to the circle; when the point (x', y') is within the circle, this difference is the square of half the least chord through the point.

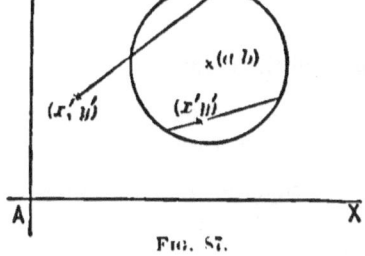

Fig. 57.

Let S represent the left-hand member of the equation $x^2 + y^2 - 2ax - 2by + c = 0$. Then $S = 0$ is the equation of the circle, and S evaluated for the coordinates of any point (x, y) is the power of that point with respect to the circle.

Let $S_1 = 0$ and $S_2 = 0$ represent two given circles. $S_1 = S_2$ is the equation of the locus of the points whose powers with respect to $S_1 = 0$ and $S_2 = 0$ are equal. This equation, which may be written $S_1 - S_2 = 0$, represents a straight line called the radical axis of the two circles. The radical axis of two circles is their common chord.

If three circles are given, $S_1 = 0$, $S_2 = 0$, $S_3 = 0$, the radical axes of these circles taken two and two are $S_1 - S_2 = 0$, $S_2 - S_3 = 0$, $S_3 - S_1 = 0$. The sum of these three equations is identically zero, showing that the radical axes of three circles taken two and two pass through a common point. This point is called the radical center of the three circles.

Problems. — 1. Find the locus of the points from which tangents to the circles $x^2 + y^2 + 4x - 8y = 5$, $x^2 + y^2 - 6x = 7$ are equal.

2. Find the point from which tangents drawn to the three circles $x^2 + y^2 - 2x = 8$, $x^2 + y^2 + 4y = 12$, $x^2 + y^2 + 4x + 8y = 5$ are equal.

3. Find the length of the tangent from $(-3, 2)$ to the circle
$$(x - 7)^2 + (y - 10)^2 = 9.$$

4. Find the length of the tangent from $(10, 15)$ to the circle
$$x^2 + y^2 - 4x + 6y = 12.$$

5. Find the length of the shortest chord of the circle
$$x^2 + y^2 - 6x + 4y = 3$$
through the point $(-4, 3)$.

6. Find the equation of the radical axis of $x^2 + y^2 + 5x - 7y = 15$, $x^2 + y^2 - 3x + 8y = 10$.

7. Find the radical center of $x^2 + y^2 - 3x = 5$, $x^2 + y^2 - 4x + y = 8$, $x^2 + y^2 + 7y = 9$.

8. Find the point of intersection of the three common chords of the circles $x^2 + y^2 - 4x - 2y = 9$, $x^2 + y^2 + 2x + 2y = 11$, $x^2 + y^2 - 6x + 4y = 17$ taken in pairs.

ART. 54. — COAXAL SYSTEMS

Let $S_1 = 0$ and $S_2 = 0$ represent two circles. Then
$$S_1 - kS_2 = 0,$$
for all values of the parameter k, represents a circle through the intersections of $S_1 = 0$, $S_2 = 0$. The equation $S_1 - kS_2 = 0$, interpreted geometrically, gives the proposition, the locus of all the points whose powers with respect to two circles $S_1 = 0$, $S_2 = 0$ are in a constant ratio is a circle through the points of intersection of the given circles.

$S_1 - kS_2 = 0$, by assigning to k all possible values, represents the entire system of circles such that the radical axis of any pair of circles of the system is the radical axis of $S_1 = 0$ and $S_2 = 0$.

If the parameters of $S_1 = 0$ and $S_2 = 0$ are a', b', c' and a'', b'', c'' respectively, the parameters of $S_1 - kS_2 = 0$ are
$$\frac{a' - ka''}{1 - k}, \quad \frac{b' - kb''}{1 - k}, \quad \frac{c' - kc''}{1 - k}.$$

Let $S = 0$ represent a circle, $L = 0$ a straight line. Then $S - kL = 0$ represents the system of circles through the points of intersection of circle and line. The common radical axis of this system of circles is the line $L = 0$.

Circles having a common radical axis are called a coaxal system of circles.

Problems. — 1. Write the equation of the system of circles through the points of intersection of $x^2 + y^2 - 2x + 6y = 10$ and $x^2 + y^2 - 4y = 8$.

2. Find the equation of the circle through the points of intersection of $x^2 + y^2 - 2x + 6y = 0$, $x^2 + y^2 - 4y = 8$, and the point $(4, -2)$.

3. Find the equation of the circle through the points of intersection of $x^2 + y^2 + 10y = 6$, $\frac{2}{3}x - \frac{1}{2}y = 3$, and the point $(4, 5)$.

4. Find the equation of the locus of all the points which have equal powers with respect to all circles of the coaxal system determined by the circles $x^2 + y^2 - 3x + 7y = 15$ and $x^2 + y^2 + 5x - 4y = 12$.

Art. 55.—Orthogonal Systems

Two circles
$$x^2 + y^2 - 2a'x - 2b'y + c' = 0, \quad x^2 + y^2 - 2a''x - 2b''y + c'' = 0$$
intersect at right angles when the square of the distance between their centers equals the sum of the squares of their radii; that is, when

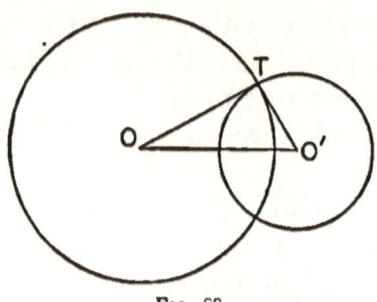

Fig. 88.

$$(a'-a'')^2 + (b'-b'')^2$$
$$= a'^2 + b'^2 - c' + a''^2 + b''^2 - c'',$$
or $\quad 2a'a'' + 2b'b'' - c' - c'' = 0.$

If the circle (a_1, b_1, c_1) cuts each of the two circles (a', b', c'), (a'', b'', c'') orthogonally, it cuts every one of the circles
$$\left(\frac{a' - ka''}{1-k}, \frac{b' - kb''}{1-k}, \frac{c' - kc''}{1-k} \right)$$
of the coaxal system orthogonally. For the hypothesis is expressed by the equations
$$2a'a_1 + 2b'b_1 - c' - c_1 = 0, \quad 2a''a_1 + 2b''b_1 - c'' - c_1 = 0;$$
the conclusion by the equation
$$2\frac{a' - ka''}{1-k}a_1 + 2\frac{b' - kb''}{1-k}b_1 - \frac{c' - kc''}{1-k} - c_1 = 0,$$
which is a direct consequence of the equations of the hypothesis.

The condition that the circle (a_1, b_1, c_1) cuts the circles (a', b', c') and (a'', b'', c'') orthogonally, is expressed by two equations between the three parameters a_1, b_1, c_1. These equations have an infinite number of solutions, showing that an infinite number of circles can be drawn, cutting the given circles orthogonally.

Let a_1, b_1, c_1 and a_2, b_2, c_2 be the parameters of any two circles $S_1 = 0$, $S_2 = 0$ cutting $S' = 0$ and $S'' = 0$ orthogonally. Then all circles of the coaxal system $S_1 - k_1 S_2 = 0$ cut orthogonally all circles of the coaxal system $S' - k'S'' = 0$. For the equations

$$2a'a_1 + 2b'b_1 - c' - c_1 = 0, \qquad (1)$$
$$2a''a_1 + 2b''b_1 - c'' - c_1 = 0, \qquad (2)$$
$$2a'a_2 + 2b'b_2 - c' - c_2 = 0, \qquad (3)$$
$$2a''a_2 + 2b''b_2 - c'' - c_2 = 0, \qquad (4)$$

have as consequence

$$2\frac{a'-k'a''}{1-k'}\frac{a_1-k_1 a_2}{1-k_1} + 2\frac{b'-k'b''}{1-k'}\frac{b_1-k_1 b_2}{1-k_1} - \frac{c'-k'c''}{1-k'} - \frac{c_1-k_1 c_2}{1-k_1} = 0.$$

Subtracting (2) from (1), and (3) from (1), there results

$$2(a'-a'')a_1 + 2(b'-b'')b_1 - c' + c'' = 0, \qquad (5)$$
$$2(a_1-a_2)a' + 2(b_1-b_2)b' - c_1 + c_2 = 0. \qquad (6)$$

Equation (5) shows that the centers of the orthogonal system $S_1 - k_1 S_2 = 0$ lie in the radical axis of the system $S' - k'S'' = 0$; equation (6) shows that the centers of the system $S' - k'S'' = 0$ lie in the radical axis of the system $S_1 - k_1 S_2 = 0$.

Take for X-axis the line of centers of the system $S' - k'S''$, for Y-axis the radical axis of this system. Then the equation of any circle of the system becomes

$$x^2 + y^2 - 2a'x + c' = 0. \quad (\alpha)$$

Since by hypothesis

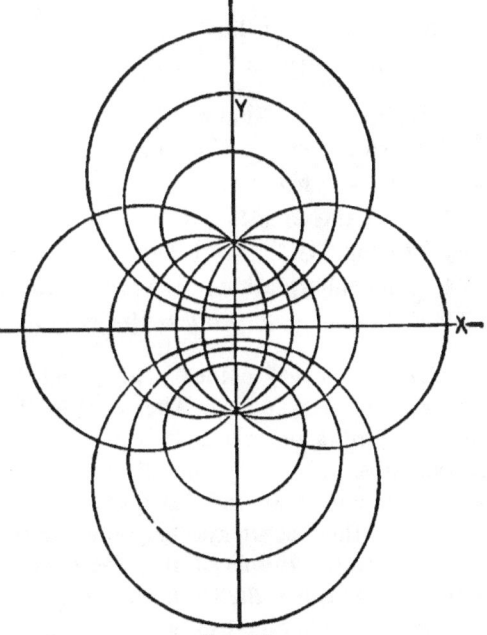

Fig. 69.

the power of $(0, y')$ is the same for all circles of the system, c' must be a fixed constant. In like manner it is found that the equations of the orthogonal system $S_1 - k_1 S_2 = 0$ have the form
$$x^2 + y^2 - 2b_1 y + c_1 = 0, \qquad (\beta)$$
where c_1 is a fixed constant. The condition for the orthogonal intersection of two circles when applied to (α) and (β) becomes $c_1 = -c'$. Hence the equations of two orthogonal systems of circles, when the radical axes of the systems are taken as reference axes, are
$$x^2 + y^2 - 2a'x + c' = 0, \qquad x^2 + y^2 - b'y - c' = 0,$$
where a' and b' are parameters, c' a fixed constant.

The radii of the circles of the two orthogonal systems are given by the equations $r^2 = a'^2 - c'$, $r'^2 = b'^2 + c'$ respectively. When r and r' become zero the circles become points, called the point circles of the system. In every case one of the orthogonal systems has a pair of real, the other a pair of imaginary, point circles.*

Problems. — 1. Find the equation of the locus of the centers of the circles which cut orthogonally the circles $x^2 + y^2 - 4x + 6y = 15$, $x^2 + y^2 + 5x - 8y = 20$.

2. Find the equation of the circle through the point $(2, -3)$ and cutting orthogonally the circles $x^2 + y^2 + 5x - 7y = 18$, $x^2 + y^2 - 2x - 4y = 12$.

3. Find the equation of the circle cutting orthogonally $x^2 + y^2 - 10x = 9$, $x^2 + y^2 = 25$, $x^2 + y^2 - 8y = 16$.

---

* Through every point of the plane there passes one circle of each of the orthogonal systems. The point in the plane is determined by giving the two circles on which it lies. This leads to a system of bicircular coordinates.

If heat enters an infinite plane disc at one point at a uniform rate, and leaves the disc at another point at the same uniform rate, when the temperature conditions of the disc have become permanent, the lines of equal temperature, the isothermal lines, and the lines of flow of heat are systems of orthogonal circles. The points where the heat enters and leaves the disc are the point circles of the isothermal system.

4. Find the equation of the system of circles cutting orthogonally the coaxal system determined by $x^2+y^2+4x+6y=15$, $x^2+y^2+2x-8y=12$.

5. Write the equation of the two orthogonal systems of circles whose real point circles are $(0, 4)$, $(0, -4)$.

Art. 56.—Tangents to Circles

The equation of a tangent to the circle $x^2+y^2=r^2$ at the point (x_0, y_0) of the circumference is $xx_0+yy_0=r^2$.

Let (x_1, y_1) be any point in the plane of the circle $x^2+y^2=r^2$, (x', y'), (x'', y''), the points of contact of tangents from (x_1, y_1) to the circle. Then (x_1, y_1) must lie in each of the lines $xx'+yy'=r^2$, $xx''+yy''=r^2$; that is, $x_1x'+y_1y'=r^2$, and $x_1x''+y_1y''=r^2$. Hence the equation of the chord of contact is $xx_1+yy_1=r^2$.

The distance from the center of the circle to the chord of contact is $\dfrac{r^2}{(x_1^2+y_1^2)^{\frac{1}{2}}}$, which is less than, equal to, or greater than r, according as the point (x_1, y_1) lies without the circumference, on the circumference, or within the circumference. In the first case the points of contact of the tangents from (x_1, y_1) to the circle are real and distinct, in the second case real and coincident, in the third case imaginary. In all cases the chord of contact is real.

In the equation $y=mx+n$ let m be a fixed constant, n a parameter. The equation represents a system of parallel straight lines. The value of n is to be determined so that the line represented by $y=mx+n$ is tangent to the circle $x^2+y^2=r^2$. The line is tangent to the circle when the perpendicular from the center of the circle to the line equals the radius; that is, when $\dfrac{n}{\pm\sqrt{1+m^2}}=r$, $n=\pm r\sqrt{1+m^2}$. Therefore, the equations of tangents to $x^2+y^2=r^2$ parallel to $y=mx+n$ are $y=mx\pm r\sqrt{1+m^2}$.

Problems. — 1. Find the equations of the tangents to $x^2 + y^2 = 25$ at $x = 3$.

2. Find the chord of contact of tangents from $(2, -3)$ to $x^2 + y^2 = 1$.

3. Find the points of contact of tangents from $(5, 7)$ to $x^2 + y^2 = 9$.

4. Find the equations of tangents to $x^2 + y^2 = 16$, making angles of $45°$ with the X-axis.

5. Find the equations of tangents to $x^2 + y^2 = 25$ parallel to $y = 3x + 5$.

6. Find the equations of tangents to $x^2 + y^2 = 25$ perpendicular to $y = 3x + 5$.

7. Find the slopes of tangents to $x^2 + y^2 = 9$ through $(4, 5)$.

8. Find the equations of the tangents to $x^2 + y^2 = 16$ through $(5, 7)$.

9. The chord of contact of a pair of tangents to $x^2 + y^2 = 25$ is $2x + 3y = 5$. Find the intersection of the tangents.

10. Find equation of tangent to $(x - a)^2 + (y - b)^2 = r^2$ at (x', y') of circumference.

Art. 57. — Poles and Polars

Since it is awkward to speak of the chord of contact or the point of intersection of a pair of imaginary tangents, the point (x_1, y_1) is called the pole of the straight line $xx_1 + yy_1 = r^2$ with respect to the circle $x^2 + y^2 = r^2$, and $xx_1 + yy_1 = r^2$ is called the

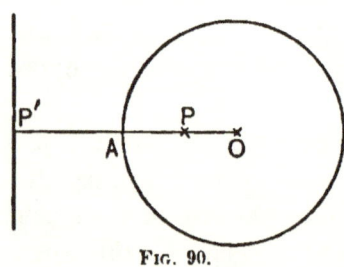

Fig. 90.

polar of the point (x_1, y_1). (x_1, y_1), which may be any point of the plane, determines uniquely the line $xx_1 + yy_1 = r^2$; and conversely, $xx_1 + yy_1 = r^2$, which may be any straight line of the plane, determines uniquely the point (x_1, y_1). The relation between pole and polar therefore establishes a one-to-one correspondence between the points of the plane and the straight lines of the plane.

The polar of (x_1, y_1) with respect to $x^2 + y^2 = r^2$ is $xx_1 + yy_1 = r^2$, the line through the center of the circle and (x_1, y_1) is $y = \dfrac{y_1}{x_1} x$. Hence the line through the pole and the center is perpen-

PROPERTIES OF THE CIRCLE

dicular to the polar, and the angle included by lines from the center to any two points equals the angle included by the polars of the two points.

The distance from the center of the circle $x^2 + y^2 = r^2$ to the polar of (x_1, y_1) is $\dfrac{r^2}{(x_1^2 + y_1^2)^{\frac{1}{2}}}$; that is, the radius is the geometric mean between the distances of the center from pole and polar.

The polar of (x_1, y_1), with respect to the circle $x^2 + y^2 = r^2$, is constructed geometrically by drawing a perpendicular to the line joining (x_1, y_1) and the center of the circle at the point whose distance from the center is the third proportional to the distance from (x_1, y_1) to the center and the radius of the circle. The pole of any line, with respect to the circle $x^2 + y^2 = r^2$, is constructed geometrically by laying off from the center on the perpendicular from the center to the line the third proportional to distance from center to line and the radius of the circle.

The polar of (x_1, y_1) with respect to the circle $x^2 + y^2 = r^2$ is $xx_1 + yy_1 = r^2$, the polar of (x_2, y_2) is $xx_2 + yy_2 = r^2$. The condition which causes (x_1, y_1) to lie in the polar of (x_2, y_2) is $x_1x_2 + y_1y_2 = r^2$; this is also the condition which causes the polar of (x_1, y_1) to contain (x_2, y_2). Hence the polars of all points in a straight line pass through the pole of the line, and the poles of all lines through a point lie in the polar of that point.

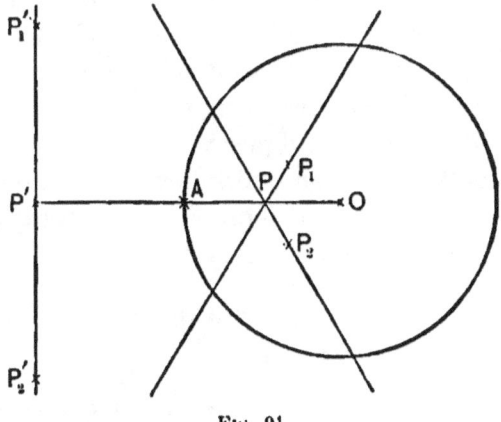

Fig. 91.

Problems. — 1. Write the equation of the polar of (2, 3) with respect to $x^2 + y^2 = 16$.

2. Find the point whose polar with respect to $x^2 + y^2 = 9$ is $8x + 7y = 18$.

3. Find distance from center of circle $x^2 + y^2 = 25$ to polar of (3, 4).

4. Find equation of polar of (x', y') with respect to circle $(x - a)^2 + (y - b)^2 = r^2$.

5. Find polar of (0, 0) with respect to $x^2 + y^2 = r^2$.

Art. 58. — Reciprocal Figures

If a geometric figure is generated by the continuous motion of a point, the polar of the generating point takes consecutive positions enveloping a geometric figure. To every point in the first figure there corresponds a tangent to the second figure; to points of the first figure in a straight line there correspond tangents to the second figure through a point; to a multiple point of the first figure there corresponds a multiple tangent in the second. If two points of intersection of a secant of the first figure become coincident, in which case the secant becomes a tangent, the pole of the secant at the same time must become the point of intersection of two consecutive tangents of the second figure, that is a point of the second figure. Hence the first figure is also the envelope of the polars of the points of the second figure. For this reason these figures are called reciprocal figures. Reciprocation leads to the principle of duality in geometry.*

Problems. — 1. To find the reciprocal of the circle C with respect to the circle O, $x^2 + y^2 = r^2$.

---

* The principle of duality was developed by Poncelet (1822) and Gergonne (1817-18) as a consequence of reciprocation, independently of reciprocation by Möbius and Gergonne.

PROPERTIES OF THE CIRCLE

The line $HH'(xx_1 + yy_1 = r^2)$ is the polar of the center $C(x_1, y_1)$ with respect to the circle O; $p(x_2, y_2)$ is the pole of any tangent $PT(xx_2 + yy_2 = r^2)$ to the circle C. Then

$$OC = (x_1^2 + y_1^2)^{\frac{1}{2}},$$

$$pK = \frac{x_1 x_2 + y_1 y_2 - r^2}{(x_1^2 + y_1^2)^{\frac{1}{2}}},$$

$$CP = \frac{x_1 x_2 + y_1 y_2 - r^2}{(x_2^2 + y_2^2)^{\frac{1}{2}}},$$

$$Op = (x_2^2 + y_2^2)^{\frac{1}{2}}.$$

Hence

$$OC \cdot pK = CP \cdot Op,$$

or

$$\frac{Op}{pK} = \frac{OC}{CP}.$$

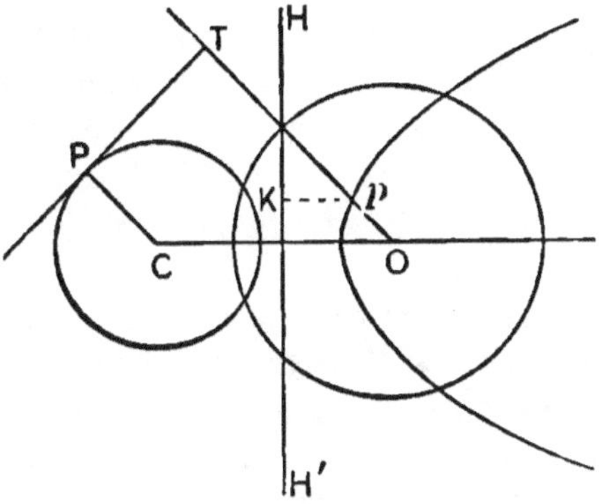

Fig. 92.

$\frac{OC}{CP}$ is constant, and therefore p must generate a conic section whose focus is O, directrix HH', eccentricity $\frac{OC}{CP}$. This conic section is an ellipse when O is within the circumference of the circle C, a parabola when O is on the circumference, an hyperbola when O is without the circumference.

2. Find the reciprocal of a given triangle.

Call the vertices of the given triangle $A(x_1, y_1)$, $B(x_2, y_2)$, $C(x_3, y_3)$. The polars of these vertices with respect to $x^2 + y^2 = r^2$ are

$$bc(xx_1 + yy_1 = r^2),$$
$$ac(xx_2 + yy_2 = r^2),$$
$$ab(xx_3 + yy_3 = r^2).$$

Triangles such that the vertices of the one are the poles of the sides of the other are called conjugate triangles. The conjugate triangle of the triangle circumscribed about a circle with respect to that circle is the triangle formed by joining the points of contact.

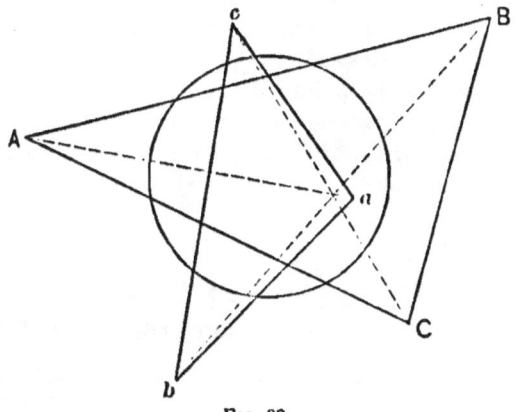

Fig. 93.

3. The straight lines joining the corresponding vertices of a pair of conjugate triangles intersect in a common point.

The equations of the lines through the corresponding vertices are

Aa, $(x_1x_3 + y_1y_3 - r^2)(xx_2 + yy_2 - r^2)$
$\quad - (x_1x_2 + y_1y_2 - r^2)(xx_3 + yy_3 - r^2) = 0$;

Bb, $(x_1x_2 + y_1y_2 - r^2)(xx_3 + yy_3 - r^2)$
$\quad - (x_2x_3 + y_2y_3 - r^2)(xx_1 + yy_1 - r^2) = 0$;

Cc, $(x_2x_3 + y_2y_3 - r^2)(xx_1 + yy_1 - r^2)$
$\quad - (x_1x_3 + y_1y_3 - r^2)(xx_2 + yy_2 - r^2) = 0.$

The sum of these equations is identically zero, therefore the lines Aa, Bb, Cc, pass through a common point.

4. Show that if a triangle is circumscribed about a circle the straight lines joining the vertices with the points of contact of the opposite sides pass through a common point.

5. Reciprocate problem 3.

The figure formed by the conjugate triangles ABC, abc is its own reciprocal. The poles of the lines joining the corresponding vertices of ABC and abc are the points of intersection of the corresponding sides of ABC and abc. Hence the reciprocal of problem 3 is, the points of intersection of the corresponding sides of a pair of conjugate triangles lie in a straight line.

6. Reciprocate problem 4.

The reciprocal of the circle is a conic section, the reciprocals of the points of contact of the sides of the triangle are tangents of the conic section, the reciprocals of the vertices of the triangle are the chords of the conic section joining the points of tangency, hence the poles of the lines from the vertices to the points of contact of the opposite sides in the given figure are the points of intersection of the sides of the triangle inscribed in the conic section with the tangents to the conic section at the opposite vertices of the triangle. These three points of intersection must lie in a straight line.

ART. 59. — INVERSION*

Let $P'(x_1, y_1)$ be any point in the plane of the circle $x^2 + y^2 = r^2$, $P(x, y)$ the intersection of the polar of P', (1) $xx_1 + yy_1 = r^2$, and

---

* The value of inversion in geometric investigation was shown by Plücker in 1831. The value of inversion in the theory of potential was shown by Lord Kelvin in 1845.

PROPERTIES OF THE CIRCLE   107

the diameter through P'', (2) $y = \frac{y_1}{x_1}x$. Then $OP \cdot OP'' = r^2$, that is $(r - PA)(r + P''A) = r^2$, whence $\frac{1}{PA} - \frac{1}{P''A} = \frac{1}{r}$. When r becomes infinite the circle becomes a straight line, $PA = P''A$, and P and P'' become symmetrical points with respect to the line. P is said to be obtained from P'' by inversion, by the transformation by reciprocal radii vectors, or by symmetry with respect to the circle. This transformation establishes a one-to-one correspondence between the points within the circle and the points without the circle.

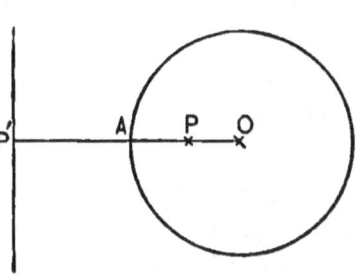

Fig. 94.

The coordinates of P are obtained in terms of the coordinates of P'' by making (1) and (2) simultaneous and solving for x and y. There results $x = \dfrac{r^2 x_1}{x_1^2 + y_1^2}$, $y = \dfrac{r^2 y_1}{x_1^2 + y_1^2}$.

Similarly, $\qquad x_1 = \dfrac{r^2 x}{x^2 + y^2}, \quad y_1 = \dfrac{r^2 y}{x^2 + y^2}.$

If the point (x_1, y_1) describes a circle
$$x_1^2 + y_1^2 - 2ax_1 - 2by_1 + c = 0,$$
the inverse point (x, y) traces a curve whose equation is
$$\frac{r^4 x^2 + r^4 y^2}{(x^2 + y^2)^2} - \frac{2ar^2 x}{x^2 + y^2} - \frac{2br^2 y}{x^2 + y^2} + c = 0,$$
which reduces to
$$x^2 + y^2 - \frac{2ar^2}{c}x - \frac{2br^2}{c}y + \frac{r^4}{c} = 0,$$
the equation of a circle. Hence inversion transforms the circle (a, b, c) into the circle $\left(\dfrac{ar^2}{c}, \dfrac{br^2}{c}, \dfrac{r^4}{c}\right)$. Calling the radius of the

given circle R, the radius of the transformed circle R',

$$R'^2 = \frac{a^2 r^4}{c^2} + \frac{b^2 r^4}{c^2} - \frac{r^4}{c} = \frac{r^4}{c^2}(a^2 + b^2 - c).$$

That is, $$R' = \frac{r^2}{c} R.$$

When $c = 0$, $R' = \infty$; that is, the transformed circle becomes a straight line. $c = 0$ is the condition which causes the center of the inversion circle, which has been taken at the origin of coordinates, to lie in the circumference of the given circle (a, b, c).

The inverse of a geometric figure may be constructed mechanically by means of an apparatus called Peaucellier's inversor. The apparatus consists of six rods, four of equal length b forming a rhombus, and two others of equal length a connecting diagonally opposite vertices of the rhombus with a fixed point O. The rods are fastened together by pins so as to allow perfect freedom of rotation about the pins. If P is made to follow a given curve, P' traces the inverse, the center of inversion being O and the radius of inversion $(a^2 + b^2)^{\frac{1}{2}}$. For

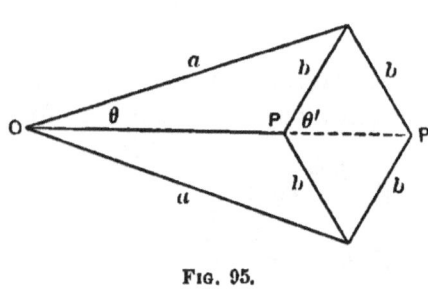

Fig. 95.

$OP = a \cos \theta - b \cos \theta'$, $OP' = a \cos \theta + b \cos \theta'$, $a \sin \theta = b \sin \theta'$.
Hence $OP \cdot OP' = a^2 \cos^2 \theta - b^2 \cos^2 \theta'$, $a^2 \sin^2 \theta - b^2 \sin^2 \theta' = 0$,
and by addition $OP \cdot OP' = a^2 - b^2$.

If the point P describes the circumference of a circle passing through O, P' must move in a straight line. Therefore the inversor transforms the circular motion of P about O', midway between O and P, as center, into the rectilinear motion of P'.

The cosine of the angle between two circles (a, b, c), (a', b', c') is found from the equation

$$(a - a')^2 + (b - b')^2 = r^2 + r'^2 - 2rr'\cos\theta$$

to be
$$\cos\theta = \frac{2aa' + 2bb' - c - c'}{2\sqrt{a^2 + b^2 - c}\sqrt{a'^2 + b'^2 - c'}}.$$

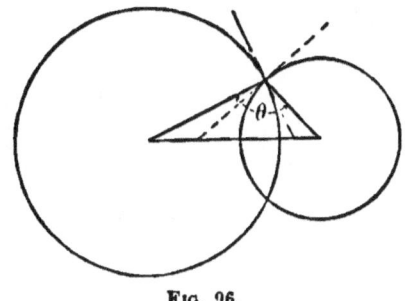

Fig. 96.

The circles obtained by inverting the given circles are

$$\left(\frac{ar^2}{c}, \frac{br^2}{c}, \frac{r^4}{c}\right), \left(\frac{a'r^2}{c'}, \frac{b'r^2}{c'}, \frac{r^4}{c'}\right).$$

Calling their included angle θ',

$$\cos\theta' = \frac{\dfrac{2aa'r^4}{cc'} + \dfrac{2bb'r^4}{cc'} - \dfrac{r^4}{c} - \dfrac{r^4}{c'}}{2\sqrt{\dfrac{a^2r^4}{c^2} + \dfrac{b^2r^4}{c^2} - \dfrac{r^4}{c}}\sqrt{\dfrac{a'^2r^4}{c'^2} + \dfrac{b'^2r^4}{c'^2} - \dfrac{r^4}{c'}}},$$

which reduces to
$$\cos\theta' = \frac{2aa' + 2bb' - c - c'}{2\sqrt{a^2 + b^2 - c}\sqrt{a'^2 + b'^2 - c'}}.$$

Hence the angle between two circles is not altered by inversion. For this reason inversion is called an equiangular or conformal transformation.

If two orthogonal systems of circles are inverted, taking for center of inversion one of the points of intersection of that system of circles which has real points of intersection, one of the systems of circles transforms into a system of straight lines through a point. Hence the other system of circles must transform into a system of concentric circles whose common center is this point.

CHAPTER IX

PROPERTIES OF THE CONIC SECTIONS

Art. 60. — General Equation

A point governed in its motion by the law — the ratio of the distances from the moving point to a fixed point and to a fixed line is constant — generates a conic section. To express this definition by an equation between the coordinates of the moving point, let the moving point be (x, y), the fixed point F, the focus (m, n), the fixed line HH', the directrix $x \cos \alpha + y \sin \alpha - p = 0$. Calling the constant ratio e, the definition is expressed by the equation $PF^2 = e^2 \cdot PD^2$, which becomes

$$(m - x)^2 + (n - y)^2 = e^2(x \cos \alpha + y \sin \alpha - p)^2.$$

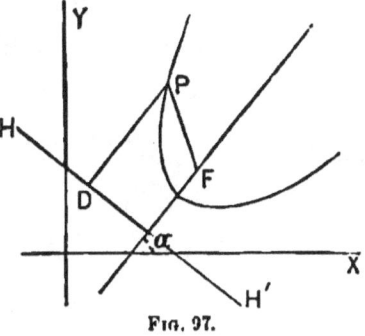

Fig. 97.

α is the angle which the axis of the conic section makes with the X-axis, p the distance from the origin to the directrix.

By assigning to m, n, e, α, p their proper values in any special case, this general equation becomes the equation of any conic section in any position whatever in the XY-plane. For example, to obtain the common equation of the ellipse, which is the equation of the ellipse referred to its axes, make $m = ae$, $n = 0$, $\alpha = 0$, $p = \dfrac{a}{e}$, $1 - e^2 = \dfrac{b^2}{a^2}$. The general equation becomes $(ae - x)^2 + y^2 = (ex - a)^2$. Expanding and collecting terms, $y^2 + (1 - e^2)x^2 = a^2(1 - e^2)$, or $\dfrac{x^2}{a^2} + \dfrac{y^2}{b^2} = 1$.

To obtain the equation of the hyperbola referred to its axis and the tangent at the left-hand vertex, make $m = a(1 + e)$,

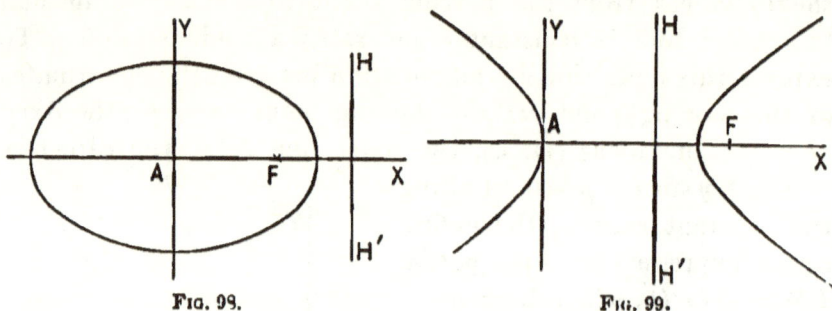

Fig. 98.   Fig. 99.

$n = 0$, $a = 0$, $p = \dfrac{a(1 + e)}{e}$, $1 - e^2 = -\dfrac{b^2}{a^2}$. The general equation becomes $(a + ae - x)^2 + y^2 = (ex - a - ae)^2$. Expanding and collecting terms, $y^2 = (1 - e^2)(2ax - x^2)$, or

$$y^2 = -\dfrac{b^2}{a^2}(2ax - x^2).$$

Problems. — From the general equation of a conic section referred to rectangular axes, obtain:

1. The common equation of the hyperbola.
2. The common equation of the parabola.
3. The equation of the ellipse referred to its axis and the tangent at the left-hand vertex.
4. The equation of the ellipse referred to its axis and the tangent at the right-hand vertex.
5. The equation of the hyperbola referred to its axis and the tangent at the right-hand vertex.
6. The equation of the parabola referred to its axis and the perpendicular to the axis through the focus.
7. The equation of the ellipse referred to its axis and the perpendicular to the axis through the focus.
8. The equation of the hyperbola referred to its axis and the directrix.

9. The equation of the parabola referred to its axis and the directrix.

10. Show that in the hyperbolas $\frac{x^2}{a^2} - \frac{y^2}{b^2} = 1$, $\frac{y^2}{b^2} - \frac{x^2}{a^2} = 1$ the transverse axis of the first is the conjugate axis of the second, and *vice versa*. Such hyperbolas are called a pair of conjugate hyperbolas.

11. Derive from the general equation of a conic section the equation of the hyperbola conjugate to $\frac{x^2}{a^2} - \frac{y^2}{b^2} = 1$.

$$m = 0, \ n = be, \ \alpha = 90°, \ p = \frac{b}{e}, \ 1 - e^2 = -\frac{a^2}{b^2}.$$

12. Show that the straight lines $y = \pm \frac{b}{a} x$ are the common asymptotes of the pair of conjugate hyperbolas $\frac{x^2}{a^2} - \frac{y^2}{b^2} = 1$, $\frac{x^2}{a^2} - \frac{y^2}{b^2} = -1$.

13. Find the equation of the ellipse focus $(-3, 2)$, eccentricity $\frac{3}{5}$, major axis 10, the axis of the ellipse making an angle of 45° with the X-axis.

14. Find the equation of the ellipse whose focal distances are 2 and 8, center $(5, 7)$, axes parallel to axes of reference.

15. Find the equation of the hyperbola whose axes are 10 and 8, center $(3, -2)$, axis of curve parallel to X-axis.

16. Find the equation of the parabola whose parameter is 6, vertex $(2, -3)$, axis of parabola parallel to X-axis.

Art. 61. — Tangents and Normals

Using the common equations of ellipse, hyperbola, and parabola, the equations of tangents to these curves at the point (x_0, y_0) of the curve are

$$\frac{xx_0}{a^2} + \frac{yy_0}{b^2} = 1, \quad \frac{xx_0}{a^2} - \frac{yy_0}{b^2} = 1,$$

$yy_0 = p(x + x_0)$, respectively. The slopes of these tangents are for the ellipse $-\frac{b^2 x_0}{a^2 y_0}$, for the hyperbola $\frac{b^2 x_0}{a^2 y_0}$, for

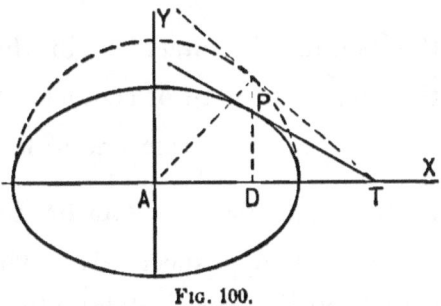

Fig. 100.

the parabola $\frac{p}{y_0}$. Calling the intercepts of the tangent on the X-axis X, on the Y-axis Y, for the ellipse $X = \frac{a^2}{x_0}$, $Y = \frac{b^2}{y_0}$, for the hyperbola $X = \frac{a^2}{x_0}$, $Y = -\frac{b^2}{y_0}$, for the parabola $X = -x_0$, $Y = \frac{1}{2} y_0$. X and Y may in each case be determined geometrically, and the tangent drawn as indicated in the figure.

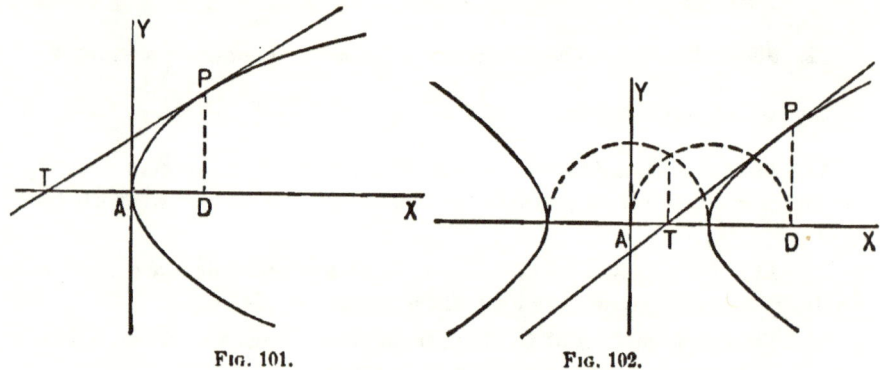

Fig. 101.　　　　Fig. 102.

Suppose the point (x_1, y_1) to be any point in the plane of the ellipse $\frac{x^2}{a^2} + \frac{y^2}{b^2} = 1$. Let (x', y'), (x'', y'') be the points of contact of tangents from (x_1, y_1) to the ellipse. Then must (x_1, y_1) lie in each of the lines $\frac{xx'}{a^2} + \frac{yy'}{b^2} = 1$, $\frac{xx''}{a^2} + \frac{yy''}{b^2} = 1$; that is, the equations $\frac{x_1 x'}{a^2} + \frac{y_1 y'}{b^2} = 1$, $\frac{x_1 x''}{a^2} + \frac{y_1 y''}{b^2} = 1$ must be true. Hence the points of contact lie in the line $\frac{xx_1}{a^2} + \frac{yy_1}{b^2} = 1$, which is therefore the chord of contact. Similarly, it is found that the points of contact from (x_1, y_1) to the hyperbola $\frac{x^2}{a^2} - \frac{y^2}{b^2} = 1$ and to the parabola $y^2 = 2px$ lie in the lines $\frac{xx_1}{a^2} - \frac{yy_1}{b^2} = 1$ and $yy_1 = p(x + x_1)$ respectively. The coordinates of the points of contact of tangents through (x_1, y_1) to a conic section are found by making the equations of the chord of contact and of the conic section simultaneous and solving for x and y.

A theory of poles and polars with respect to any conic section might be constructed entirely analogous to the theory of poles and polars with respect to the circle.

The equation $y = mx + n$, where m is a fixed constant, n a parameter, represents a system of parallel straight lines. For any value of n, the abscissas of the points of intersection of straight line and ellipse $\frac{x^2}{a^2} + \frac{y^2}{b^2} = 1$ are found by solving the equation $(b^2 + a^2 m^2) x^2 + 2 a^2 mn x + a^2 (n^2 - b^2) = 0$. These abscissas are equal, and the line $y = mx + n$ becomes a tangent to the ellipse $\frac{x^2}{a^2} + \frac{y^2}{b^2} = 1$ when $n^2 = b^2 + a^2 m^2$. Therefore $y = mx \pm (b^2 + a^2 m^2)^{\frac{1}{2}}$ are the two tangents to the ellipse whose slope is m. In like manner it is found that the tangents to the hyperbola whose slope is m are $y = mx \pm (a^2 m^2 - b^2)^{\frac{1}{2}}$; the tangent to the parabola whose slope is m is $y = mx + \frac{p}{2m}$.

The equations of the normals to ellipse, hyperbola, and parabola at the point (x_0, y_0) of the curves are $y - y_0 = \frac{a^2 y_0}{b^2 x_0}(x - x_0)$, $y - y_0 = -\frac{a^2 y_0}{b^2 x_0}(x - x_0)$, $y - y_0 = -\frac{y_0}{p}(x - x_0)$ respectively.

Problems. — 1. Find the equations of tangents to the ellipse whose axes are 8 and 6 at the points whose distance from the Y-axis is 1.

2. Find the equations of the focal tangents of ellipse, hyperbola, and parabola.

3. From the point (6, 8) tangents are drawn to the ellipse $\frac{x^2}{16} + \frac{y^2}{9} = 1$. Find the coordinates of the points of contact and the equations of the tangents.

4. At what point of the parabola $y^2 = 10x$ is the slope of the tangent $1\frac{1}{4}$?

5. On an elliptical track whose major axis is due east and west and 1 mile long, minor axis $\frac{3}{4}$ mile long, in what direction is a man traveling

when walking from west to east and $\frac{1}{8}$ mile west of the north and south line?

6. Write the equations of tangents to $\frac{x^2}{9} + \frac{y^2}{4} = 1$ making an angle $45°$ with the X-axis.

7. Write the equations of the tangents to $\frac{x^2}{9} - \frac{y^2}{4} = 1$ perpendicular to $2x - 3y = 4$.

8. Write the equation of the tangent to $y^2 = 8x$ parallel to $\frac{x}{2} + \frac{y}{3} = 1$.

9. Find the slopes of the tangents to $\frac{x^2}{9} + \frac{y^2}{4} = 1$ through the point $(4, 5)$. $y = mx + (4 + 9m^2)^{\frac{1}{2}}$ is tangent to $\frac{x^2}{9} + \frac{y^2}{4} = 1$. Since $(4, 5)$ is in the tangent, $5 = 4m + (4 + 9m^2)^{\frac{1}{2}}$. Solve for m.

10. Find the slopes of tangents to $\frac{x^2}{9} - \frac{y^2}{4} = 1$ through $(2, 3)$.

11. Find the slopes of tangents to $y^2 = 6x$ through $(-5, 4)$.

12. Find the points of contact of tangents to $y^2 = 6x$ through $(-5, 4)$.

13. Find the intercepts of normals to ellipse, hyperbola, and parabola on X-axis.

14. Find distances from focus to point of intersection of normal with axis for each of the conic sections.

15. Prove that tangents to ellipse, hyperbola, or parabola at the extremities of chords through a fixed point intersect on a fixed straight line.

16. Prove that the chords of contact of tangents to a conic section from points in a straight line pass through a common point.

17. Show that the tangent to the ellipse at any point bisects the angle made by one focal radius to the point with the prolongation of the other focal radius to the point.

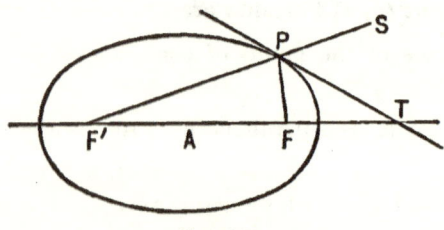

Fig. 103.

PROPERTIES OF THE CONIC SECTIONS

The ratio of the focal radii is $\dfrac{PF}{PF'} = \dfrac{a - ex_0}{a + ex_0}$. Since

$$AF = AF' = ae \text{ and } AT = \dfrac{a^2}{x_0}, \quad \dfrac{FT}{F'T} = \dfrac{\dfrac{a}{x_0}(a - ex_0)}{\dfrac{a}{x_0}(a + ex_0)}$$

Hence $\dfrac{FT}{F'T} = \dfrac{PF}{PF'}$, and PT bisects FPS.

18. In the hyperbola the tangent at any point bisects the angle included by the focal radii to the point.

19. In the parabola the tangent at any point bisects the angle included by the focal radius to and the diameter through the point.*

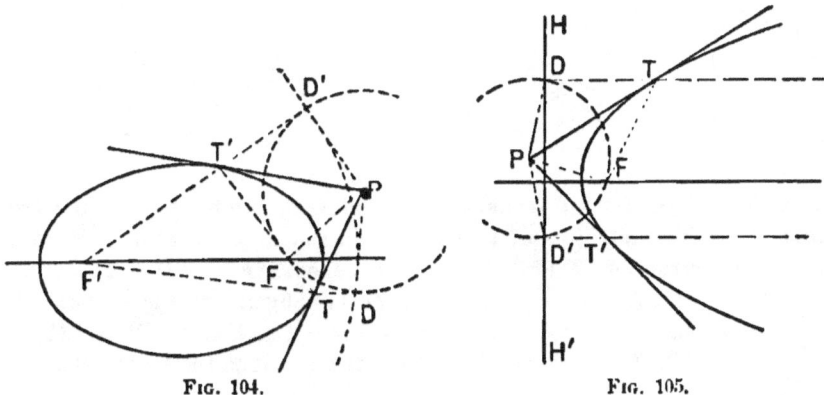

Fig. 104.   Fig. 105.

On problems 17, 18, 19 is based a simple method of drawing tangents to the conic sections through a given point. With the given point as center and radius equal to distance from given point to one focus strike

---

* Since it is true of rays of light, heat, and sound that the reflected ray and the incident ray lie on different sides of the normal and make equal angles with the normal, it follows that rays emitted from one focus of an elliptic reflector are concentrated at the other focus; that rays emitted from one focus of an hyperbola reflector proceed after reflection as if emitted from the other focus; that rays emitted from the focus of a parabolic reflector after reflection proceed in parallel lines.

It is this property of conic sections that suggested the term focus or "burning point."

off an arc. In the parabola the parallels to the axis through the intersections of this circle with the directrix determine the points of tangency. For $TF = TD$, hence the triangles TPF, TPD are equal and PT is tangent to the parabola. In ellipse and hyperbola strike off another arc with the second focus as center and radius equal to transverse axis.

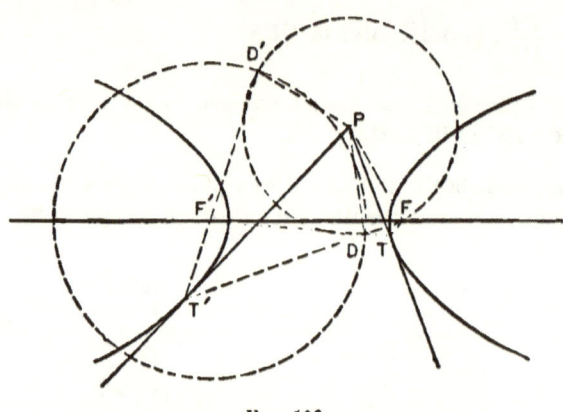

Fig. 106.

Lines joining the second focus with the points of intersection of the two arcs determine the points of tangency. In the ellipse $T'F' + T'F = 2a$, and by construction $T'F + T'D' = 2a$, hence $T'F = T'D'$. The triangles $T'PF$, $T'PD'$ are equal, and PT' is tangent to the ellipse. In the hyperbola $TF' - TF = 2a$, $TF' - TD = 2a$; hence $TD = TF$, the triangles TPD, TPF are equal, and PT is tangent to the hyperbola.

20. Show that the locus of the foot of the perpendicular from the focus of the ellipse $\dfrac{x^2}{a^2} + \dfrac{y^2}{b^2} = 1$ to the tangent is the circle described on the major axis as diameter.

The equation of the perpendicular from the focus $(ae, 0)$ to the tangents $y = mx \pm (b^2 + a^2 m^2)^{\frac{1}{2}}$ is $my + x = ae$. Make these equations simultaneous and eliminate m by squaring both equations and adding. There results $x^2 + y^2 = a^2$.

21. Show that the locus of the foot of the perpendicular from the focus of the hyperbola $\dfrac{x^2}{a^2} - \dfrac{y^2}{b^2} = 1$ to the tangent is the circle described on the transverse axis as diameter.

22. Show that the locus of the foot of the perpendicular from the focus of the parabola $y^2 = 2px$ to the tangent is the Y-axis.

PROPERTIES OF THE CONIC SECTIONS

Problems 20, 21, 22 may be used to construct the conic sections as envelopes when the focus and the vertices are known.

23. Prove that for ellipse and hyperbola the product of the perpendiculars from foci to tangent is constant and equal to b^2.

24. Prove that in the parabola the locus of the point of intersection of a line through the vertex perpendicular to a tangent with the ordinate through the point of tangency is a semi-cubic parabola.

Art. 62. — Conjugate Diameters

Let (x_0, y_0) be the point of intersection of the diameter $y = \tan\theta \cdot x$ with the ellipse $\frac{x^2}{a^2} + \frac{y^2}{b^2} = 1$, and call the angle made by the tangent to the ellipse at (x_0, y_0) with the X-axis θ'. Then

$$\tan\theta = \frac{y_0}{x_0},$$

$$\tan\theta' = -\frac{b^2 x_0}{a^2 y_0},$$

$$\tan\theta \tan\theta' = -\frac{b^2}{a^2}.$$

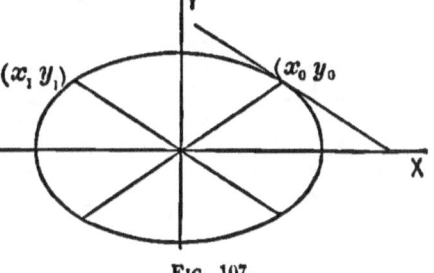

Fig. 107.

Now let (x_1, y_1) be the point of intersection of

$$y = \tan\theta' \cdot x$$

with the ellipse, and call the angle made by the tangent to the ellipse at (x_1, y_1) with the X-axis θ. Then

$$\tan\theta' = \frac{y_1}{x_1}, \quad \tan\theta = -\frac{b^2 x_1}{a^2 y_1}, \quad \tan\theta \tan\theta' = -\frac{b^2}{a^2}.$$

Hence the condition $\tan\theta \tan\theta' = -\frac{b^2}{a^2}$ causes each of the diameters of the ellipse $y = \tan\theta \cdot x$, $y = \tan\theta' \cdot x$ to be parallel to the tangent at the extremity of the other. Such diameters are called conjugate diameters of the ellipse.

The equation of the ellipse $\frac{x^2}{a^2}+\frac{y^2}{b^2}=1$ referred to a pair of conjugate diameters and in terms of the semi-conjugate diameters a' and b' is $\frac{x^2}{a'^2}+\frac{y^2}{b'^2}=1$. (See Art. 35, Prob. 39.) This equation shows that each of a pair of conjugate diameters bisects all chords parallel to the other. The axes of the ellipse are a pair of perpendicular conjugate diameters.

Let (x_0, y_0) be the point of intersection of $y = \tan\theta \cdot x$ with the hyperbola $\frac{x^2}{a^2}-\frac{y^2}{b^2}=1$, and call the angle made by the tangent to the hyperbola at (x_0, y_0) with the X-axis θ'. Then $\tan\theta = \frac{y_0}{x_0}$, $\tan\theta' = \frac{b^2 x_0}{a^2 y_0}$, $\tan\theta \tan\theta' = \frac{b^2}{a^2}$. Since $y = \frac{b}{a}x$ and

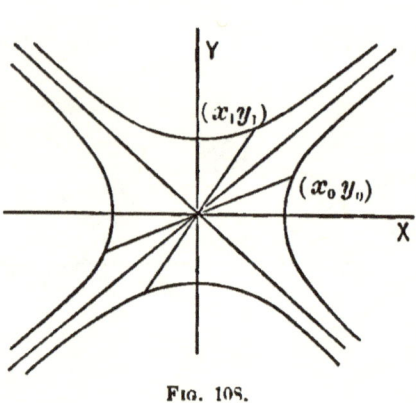

Fig. 108.

$y = -\frac{b}{a}x$ are the common asymptotes of the pair of conjugate hyperbolas $\frac{x^2}{a^2}-\frac{y^2}{b^2}=1$ and $\frac{x^2}{a^2}-\frac{y^2}{b^2}=-1$, it is evident that the condition

$$\tan\theta \tan\theta' = \frac{b^2}{a^2}$$

causes $y = \tan\theta' \cdot x$ to intersect $\frac{x^2}{a^2}-\frac{y^2}{b^2}=-1$ if $y = \tan\theta \cdot x$

intersects $\frac{x^2}{a^2}-\frac{y^2}{b^2}=1$. Now suppose (x_1, y_1) to be the point of intersection of the line $y = \tan\theta' \cdot x$ with the conjugate hyperbola $\frac{x^2}{a^2}-\frac{y^2}{b^2}=-1$, and call the angle made by the tangent to this hyperbola at (x_1, y_1) with the X-axis θ. Then $\tan\theta' = \frac{y_1}{x_1}$, $\tan\theta = \frac{b^2 x_1}{a^2 y_1}$, $\tan\theta \tan\theta' = \frac{b^2}{a^2}$. Diameters of the hyperbola satisfying the condition $\tan\theta \tan\theta' = \frac{b^2}{a^2}$ are called conjugate diameters of the hyperbola.

PROPERTIES OF THE CONIC SECTIONS 121

The equation of the hyperbola $\frac{x^2}{a^2} - \frac{y^2}{b^2} = 1$ referred to a pair of conjugate diameters, and in terms of the semi-conjugate diameters a' and b', is $\frac{x^2}{a'^2} - \frac{y^2}{b'^2} = 1$. (See Art. 35, Prob. 38.) This equation shows that chords of an hyperbola parallel to any diameter are bisected by the conjugate diameter. The axes of the hyperbola are perpendicular conjugate diameters.

The equation of the parabola referred to a diameter, and a tangent at the extremity of the diameter, is $y^2 = 2p_1 x$. (See Art. 35, Prob. 40.) This equation shows that any diameter of the parabola bisects all chords parallel to the tangent at the extremity of the diameter. The axis of the parabola is that diameter which bisects the system of parallel chords at right angles.

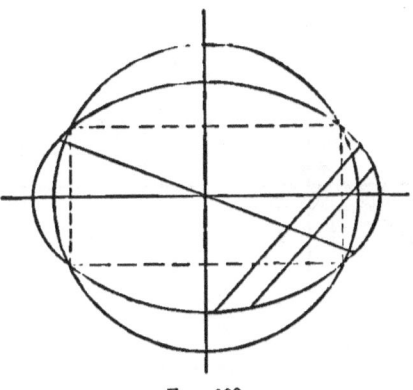

Fig. 109.

It is now possible to determine geometrically the axes, focus, and directrix of a conic section when the curve only is given. In the case of the ellipse draw any pair of parallel chords. Their bisector is a diameter of the ellipse. With the center of the ellipse as center strike off a circle intersecting the ellipse in four points. The bisectors of the two pairs of parallel chords joining the points of intersection are the axes

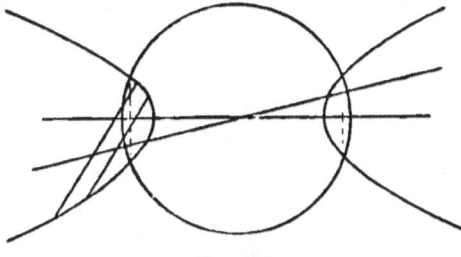

Fig. 110.

of the ellipse. An arc struck off with extremity of minor axis as center, and radius equal to semi-major axis, intersects the major axis in the foci. The directrix is perpendicular to the line of foci where the focal tangents cross this line.

In the case of the hyperbola the directions of the axes are found as for the ellipse. The focus is determined by drawing a perpendicular to any tangent at the point of intersection of this tangent with the circumference on the transverse axis. Drawing the focal tangents determines the directrix. The conjugate axis is limited by the arc struck off with vertex as center and radius equal to distance from focus to center.

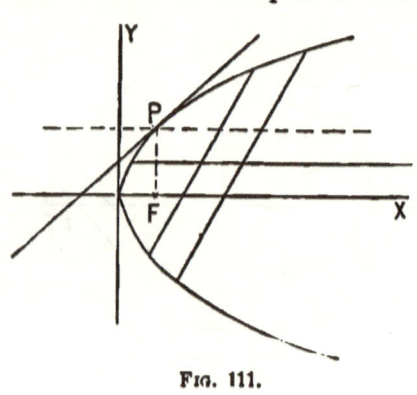

Fig. 111.

In the case of the parabola, after determining a diameter by bisecting any pair of parallel chords, and the axis by bisecting a pair of chords perpendicular to the diameter, the focus is determined by the property that the tangent bisects the angle included by diameter and focal radius to point of tangency.

Art. 63. — Supplementary Chords

Chords from any point of ellipse or hyperbola to the extremities of the transverse axis are called supplementary. Let (x', y') be any point of the ellipse $\dfrac{x^2}{a^2} + \dfrac{y^2}{b^2} = 1$. The equations of lines through (x', y'), $(a, 0)$ and (x', y'), $(-a, 0)$ are

$$y = \frac{y'}{x' - a}(x - a),$$

$$y = \frac{y'}{x' + a}(x + a).$$

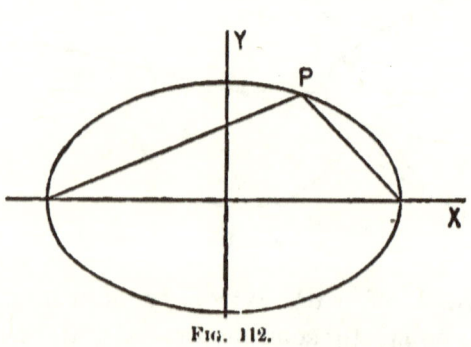

Fig. 112.

PROPERTIES OF THE CONIC SECTIONS

Calling the angles made by the supplementary chords with the X-axis ϕ and ϕ', $\tan\phi \tan\phi' = \dfrac{y'^2}{x'^2 - a^2}$. From the equation of the ellipse, $y'^2 = \dfrac{b^2}{a^2}(a^2 - x'^2)$. Hence $\tan\phi \tan\phi' = -\dfrac{b^2}{a^2}$. In like manner for the hyperbola $\dfrac{x^2}{a^2} - \dfrac{y^2}{b^2} = 1$, $\tan\phi \tan\phi' = \dfrac{b^2}{a^2}$.

$y = \tan\theta \cdot x$ and $y = \tan\theta' \cdot x$ are a pair of conjugate diameters of the ellipse $\dfrac{x^2}{a^2} + \dfrac{y^2}{b^2} = 1$ when $\tan\theta \cdot \tan\theta' = -\dfrac{b^2}{a^2}$. Hence $\tan\theta \cdot \tan\theta' = \tan\phi \cdot \tan\phi'$, from which it follows that if one of a pair of supplementary chords is parallel to a diameter the other chord is parallel to the conjugate diameter. This proposition is demonstrated for the hyperbola in the same manner.

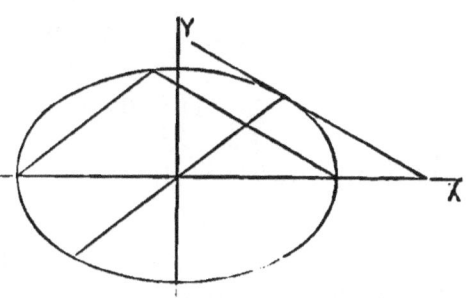

Fig. 113.

On this proposition are based simple methods of drawing tangents to ellipse or hyperbola, either through a point of the curve or parallel to a given line. To draw a tangent to the ellipse at any point P, draw a diameter through P, a supplementary chord parallel to this diameter, and the line through P parallel to the other supplementary chord is the tangent.

To draw a tangent to the hyperbola parallel to a given straight line, draw one supplementary chord parallel to the given line, and the diameter parallel to the other supplementary chord determines the points of tangency.

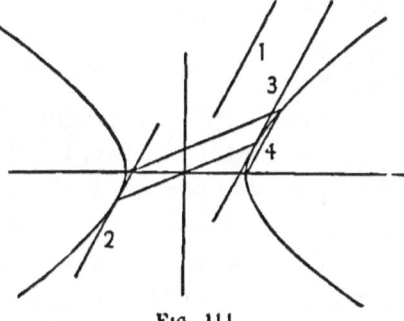

Fig. 114.

To draw a pair of conjugate diameters of an ellipse, including a given angle, construct on the major axis of the ellipse a circular segment containing the given angle. From the point of intersection of the arc of the segment and the ellipse draw a pair of supplementary chords. The diameters parallel to these chords are the required diameters.

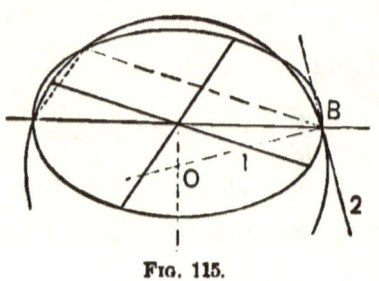

Fig. 115.

Art. 64. — Parameters

Since $\frac{x^2}{a^2} + \frac{y^2}{b^2} = 1$ is the equation of an ellipse referred to any pair of conjugate diameters, it is readily shown that the squares of ordinates to any diameter of the ellipse are in the ratio of the rectangles of the segments into which these ordinates divide the diameter. The same proposition is true of the hyperbola.

Taking the pair of perpendicular conjugate diameters of the ellipse as reference axes and the points (ae, p), $(0, b)$, the proposition leads to the proportion $\frac{p^2}{b^2} = \frac{a^2(1-e^2)}{a^2}$, whence $\frac{2p}{2b} = \frac{2b}{2a}$; that is, the parameter to the transverse axis of the ellipse is a fourth proportional to the transverse and conjugate axes. Generalizing this result, the parameter to any diameter of ellipse or hyperbola is the fourth proportional to that diameter and its conjugate.

In the common equation of the parabola, $y^2 = 2px$, the parameter $2p$ is the fourth proportional to any abscissa and its corresponding ordinate. Generalizing this definition, the parameter to any diameter of the parabola is the fourth proportional to any abscissa and its corresponding ordinate with respect to this diameter.

PROPERTIES OF THE CONIC SECTIONS

When (m, n) on the parabola $y^2 = 2px$ is taken as origin, the diameter through (m, n) as X-axis, the tangent at (m, n) as Y-axis, the equation of the parabola takes the form

$$y_1^2 = 2p_1 x_1.$$

(See Art. 35, Prob. 40.)

$$2p_1 = \frac{2p}{\sin^2 \theta'}, \quad \tan \theta' = \frac{p}{n},$$

$$n^2 = 2pm.$$

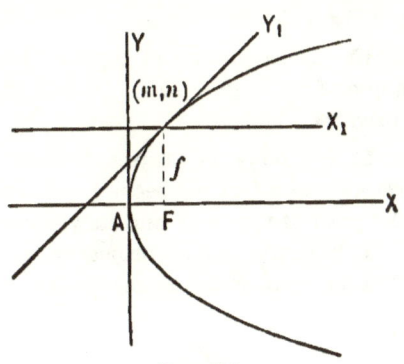

Fig. 116.

Hence $2p_1 = 4(m + \tfrac{1}{2}p)$; that is, the parameter to any diameter of a parabola is four times the focal radius of the vertex of that diameter. Calling the focal radius f, the equation of the parabola becomes $y_1^2 = 4f x_1$.

Problems. — 1. In the ellipse $\frac{x^2}{9} + \frac{y^2}{4} = 1$, find the equation of the diameter conjugate to $y = x$.

2. Find the angle between the supplementary chords of the ellipse $\frac{x^2}{a^2} + \frac{y^2}{b^2} = 1$ at the extremity of the minor axis.

3. Find the point of the ellipse $\frac{x^2}{9} + \frac{y^2}{4} = 1$ at which supplementary chords include an angle of 45°.

4. Show that the maximum angle between a pair of supplementary chords of the ellipse $\frac{x^2}{a^2} + \frac{y^2}{b^2} = 1$ is $\tan^{-1} \frac{2ab}{b^2 - a^2}$.

5. Show that a pair of conjugate diameters of an hyperbola cannot include an angle greater than 90°.

6. Construct the ellipse whose equation referred to a pair of conjugate diameters including an angle of 45° is $\frac{x^2}{9} + \frac{y^2}{4} = 1$. Find focus and directrix of this ellipse.

7. Find the equation of the hyperbola whose axes are 8 and 6 referred to a pair of conjugate diameters, of which one makes an angle of 45° with the axis of the hyperbola. Find lengths of the semi-conjugate diameters.

8. Find equation of parabola whose parameter is 8 referred to diameter through (8, 8) and tangent at this point.

9. Find the locus of the centers of chords of $\frac{x^2}{9} + \frac{y^2}{4} = 1$ parallel to $y = 2x + 5$.

10. The equation of a parabola referred to oblique axes including an angle of 60° is $y^2 = 10 x$. Sketch the parabola and construct its focus and directrix.

11. A body is projected from A in the direction AY with initial velocity of v feet per second. Gravity is the only disturbing force. Find the path of the body and its velocity at any instant.

Taking the line of projection as Y-axis and the vertical through A as X-axis, the coordinates of the body t seconds after projection are $x = \frac{1}{2} g t^2$, $y = v t$; the equation of the path of the body, found by eliminating t, is $y^2 = \frac{2 v^2}{g} x$, a parabola referred to a tangent and diameter through point of tangency. Comparing this equation with $y^2 = 4 f x$, the equation of parabola referred to tangent and diameter, $v^2 = 2 g f$; that is, the initial velocity is the velocity acquired by a body falling freely from the directrix of the parabola to the starting point.

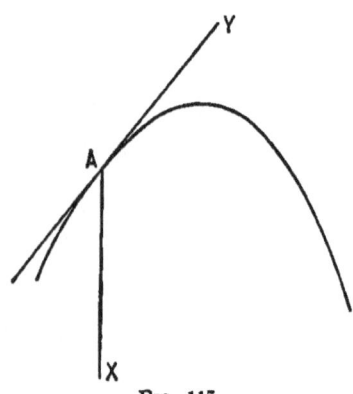

Fig. 117.

If the body is projected from any point of the parabola along the tangent to the parabola at that point, and with a velocity equal to the velocity of the body projected from A when it reaches that point, the path of the body is the path of the body projected from A. Hence it follows that the velocity of the body at any point of the parabola is the velocity acquired by a body freely falling from the directrix of the parabola to that point.

Art. 65. — The Elliptic Compass

Let $\frac{x^2}{a^2} + \frac{y^2}{b_1^2} = 1$ and $\frac{x^2}{a^2} + \frac{y^2}{b_2^2} = 1$ be two ellipses constructed on the same major diameter. Let y_1 and y_2 be ordinates corresponding to the same abscissa, then $\frac{y_1}{y_2} = \frac{b_1}{b_2}$; that is, if ellipses are constructed on the same major diameter, corresponding ordinates are to each other as the minor diameters. The circle described on the major diameter of the ellipse is a variety of the ellipse, hence the ordinate of an ellipse is to the corresponding ordinate of the circumscribed circle as the minor diameter of the ellipse is to the major diameter.

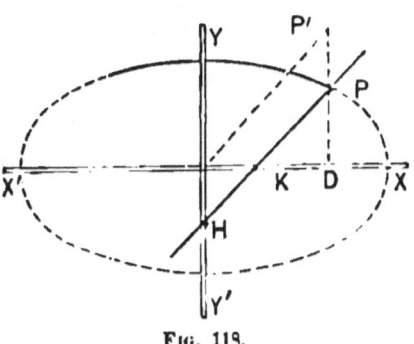

Fig. 118.

On this principle is based a convenient instrument for drawing an ellipse whose axes are given. On a rigid bar take $PH = a$, $PK = b$. Fix pins at H and K which slide in grooves in the rulers X and Y perpendicular to each other. P traces the ellipse $\dfrac{x^2}{a^2} + \dfrac{y^2}{b^2} = 1$. For $\dfrac{PD}{PD} = \dfrac{PK}{PH} = \dfrac{b}{a}$. This instrument is called the elliptic compass.

Art. 66. — Area of the Ellipse

Erect any number of perpendiculars to the major diameter of the ellipse, and beginning at the right draw through the points of intersection of these perpendiculars with the ellipse and the circumscribed circle parallels to the minor diameter. There is thus inscribed in the ellipse and in the circle a series of rectangles. The corresponding rectangles in ellipse and circle have the same base, and their altitudes are in the ratio of b to a.

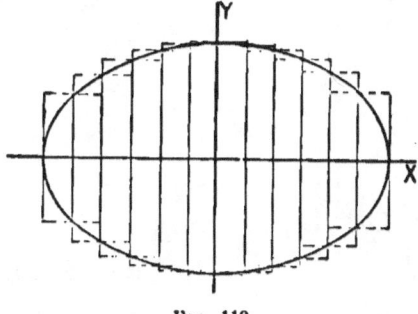

Fig. 119.

Hence the sum of the areas of the rectangles inscribed in the ellipse bears to the sum of the rectangles inscribed in the circle the ratio of b to a. By indefinitely increasing the number of rectangles, the sum of the areas of the rectangles inscribed in the ellipse approaches the area of the ellipse as its limit, and at the same time the sum of the areas of the rectangles inscribed in the circle approaches the area of the circle as its limit. At the limit therefore $\frac{\text{area of ellipse}}{\text{area of circle}} = \frac{b}{a}$, hence area of ellipse $= \frac{b}{a} \cdot \pi a^2 = \pi a b$.

Art. 67. — Eccentric Angle of Ellipse

At any point (x, y) of the ellipse $\frac{x^2}{a^2} + \frac{y^2}{b^2} = 1$ produce the ordinate to the transverse axis to meet the circumscribed circle and draw the radius of this circle to the point of meeting.

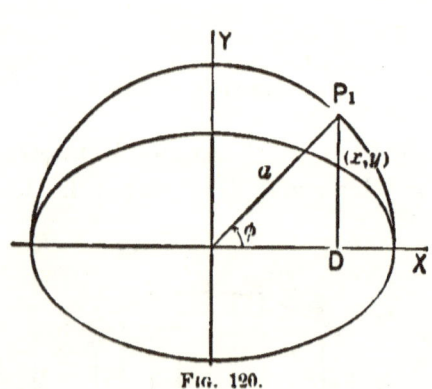

Fig. 120.

The angle ϕ made by this radius with the transverse axis of the ellipse is called the eccentric angle of the point (x, y). From the figure

$$x = a \cdot \cos \phi,$$
$$y = \frac{b}{a} \cdot P_1 D = \frac{b}{a} \cdot a \sin \phi$$
$$= b \cdot \sin \phi.$$

The coordinates of any point (x, y) of the ellipse are thus expressed in terms of the single variable ϕ.

Let AP_1 and AP_2 be a pair of conjugate diameters of the ellipse $\frac{x^2}{a^2} + \frac{y^2}{b^2} = 1$, θ and θ' the angles these diameters make with the axis of the ellipse. Then $\tan \theta \tan \theta' = -\frac{b^2}{a^2}$. Let

PROPERTIES OF THE CONIC SECTIONS 129

(x_1, y_1) be the coordinates, ϕ_1 the eccentric angle of P_1; (x_2, y_2) the coordinates, ϕ_2 the eccentric angle of P_2. Then

$$\tan \theta = \frac{y_1}{x_1} = \frac{b \sin \phi_1}{a \cos \phi_1},$$

$$\tan \theta' = \frac{y_2}{x_2} = \frac{b \sin \phi_2}{a \cos \phi_2},$$

$$\tan \theta \tan \theta'$$

$$= \frac{b^2 \sin \phi_1 \sin \phi_2}{a^2 \cos \phi_1 \cos \phi_2}$$

$$= \frac{b^2}{a^2} \tan \phi_1 \tan \phi_2$$

$$= -\frac{b^2}{a^2}.$$

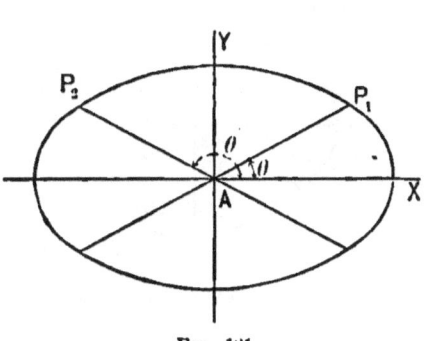

Fig. 121.

Hence $\tan \phi_1 \tan \phi_2 = -1$ and ϕ_1 and ϕ_2 differ by $90°$; that is, the eccentric angles of the extremities of a pair of conjugate diameters of the ellipse differ by $90°$.

Call the lengths of the semi-conjugate diameters a_1 and b_1. Then
$$a_1^2 = x_1^2 + y_1^2 = a^2 \cos^2 \phi_1 + b^2 \sin^2 \phi_1,$$
$$b_1^2 = a^2 \cos^2 \phi_2 + b^2 \sin^2 \phi_2 = a^2 \sin^2 \phi_1 + b^2 \cos^2 \phi_1,$$

since $\phi_2 = 90° + \phi_1$. By addition $a_1^2 + b_1^2 = a^2 + b^2$; that is, the sum of the squares of any pair of conjugate diameters of the ellipse equals the sum of the squares of the axes.

The conjugate diameters are of equal length when

$$a^2 \cos^2 \phi + b^2 \sin^2 \phi = a^2 \sin^2 \phi + b^2 \cos^2 \phi;$$

that is, when

$$\tan^2 \phi = 1, \tan \phi = \pm 1, \phi = 45° \text{ or } 135°.$$

The equations of the equal conjugate diameters are $y = \pm \frac{b}{a} x$, and their length $\sqrt{2(a^2 + b^2)}$.

The area of the parallelogram circumscribed about the

K

ellipse with its sides parallel to a pair of conjugate diameters is $4 b' \cdot AN$. The equation of the tangent to the ellipse at (x', y') is

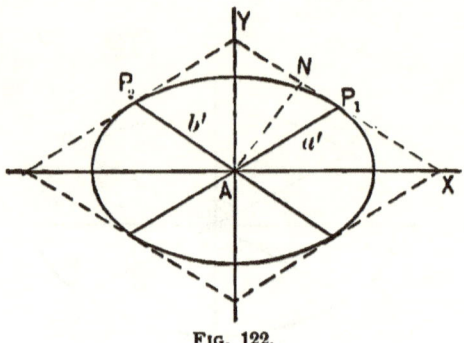

Fig. 122.

$$\frac{xx'}{a^2} + \frac{yy'}{b^2} = 1.$$

The point (x', y') is the same as $(a \cos \phi_1, b \sin \phi_1)$, and the tangent may be written

$$\frac{x \cos \phi_1}{a} + \frac{y \sin \phi_1}{b} = 1.$$

The length of the perpendicular from the origin to this tangent is
$$AN = \frac{1}{\left(\frac{\cos^2 \phi_1}{a^2} + \frac{\sin^2 \phi_1}{b^2}\right)^{\frac{1}{2}}} = \frac{ab}{b_1}.$$

Hence $4 b' \cdot AN = 4 ab$; that is, the area of the circumscribed parallelogram equals the area of the rectangle on the axes.

Art. 68. — Eccentric Angle of the Hyperbola

On the transverse axis of the hyperbola describe a circle. Through the foot of the ordinate of any point (x, y) of the hyperbola draw a tangent to this circle; the angle made by the radius to the point of tangency and the axis of the hyperbola is called the eccentric angle of the point (x, y). From the figure $x = a \cdot \sec \phi$ and, since

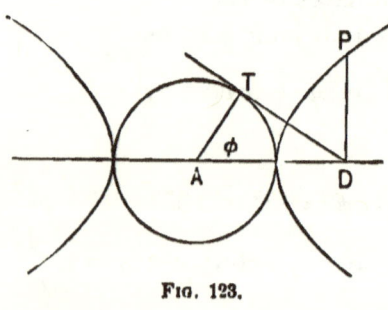

Fig. 123.

$$y^2 = \frac{b^2}{a^2}(a^2 - x^2), \; y = b \cdot \tan \phi.$$

Let AP_1 and AP_2 be a pair of conjugate diameters of the

hyperbola $\frac{x^2}{a^2} - \frac{y^2}{b^2} = 1$; (x_1, y_1) the coordinates, ϕ_1 the eccentric angle of the point P_1; θ and θ' the angles included by the conjugate diameters and the axes of the hyperbola. Then

$$\tan \theta = \frac{y_1}{x_1} = \frac{b \tan \phi_1}{a \sec \phi_1}.$$

Since $\tan \theta \tan \theta' = \frac{b^2}{a^2}$,

$$\tan \theta' = \frac{b}{a \sin \phi_1}.$$

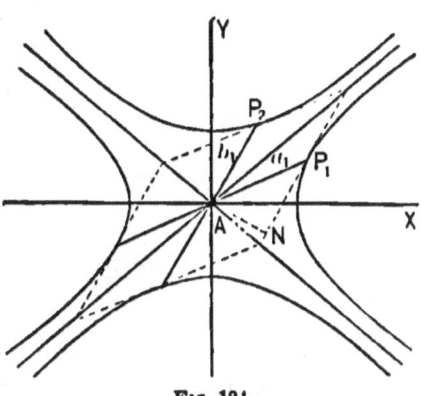

Fig. 124.

Hence the equations of the conjugate diameters are $y = \frac{b \sin \phi_1}{a} \cdot x$, $y = \frac{b}{a \sin \phi_1} \cdot x$. P_2, the point of intersection of $y = \frac{b}{a \sin \phi_1} \cdot x$ with $\frac{x^2}{a^2} - \frac{y^2}{b^2} = -1$, is $(a \tan \phi_1, b \sec \phi_1)$. Therefore $AP_1^2 = a_1^2 = a^2 \sec^2 \phi + b^2 \tan^2 \phi$, $AP_2^2 = b_1^2 = a^2 \tan^2 \phi_1 + b^2 \sec^2 \phi_1$. By subtraction $a_1^2 - b_1^2 = a^2 - b^2$; that is, the difference between the squares of any pair of conjugate diameters of the hyperbola equals the difference of the squares of the axes.

The area of the parallelogram whose sides are tangents to a pair of conjugate hyperbolas at the extremities of a pair of conjugate diameters is $4 b_1 \cdot AN$. The equation of the tangent to $\frac{x^2}{a^2} - \frac{y^2}{b^2} = 1$ at $(a \sec \phi_1, b \tan \phi_1)$ is $\frac{\sec \phi_1}{a} x - \frac{\tan \phi_1}{b} y = 1$. The perpendicular from the origin to this tangent is

$$AN = \frac{1}{\left(\frac{\sec^2 \phi_1}{a^2} + \frac{\tan^2 \phi_1}{b^2}\right)^{\frac{1}{2}}} = \frac{ab}{b_1}.$$

Hence the area of the parallelogram equals $4 ab$; that is, the area of the rectangle on the axes.

The equations of the sides of the parallelogram are

$$\frac{\sec\phi_1}{a} \cdot x - \frac{\tan\phi_1}{b} \cdot y = 1, \tag{1}$$

$$\frac{-\sec\phi_1}{a} \cdot x + \frac{\tan\phi_1}{b} \cdot y = 1, \tag{2}$$

$$\frac{\tan\phi_1}{a} \cdot x - \frac{\sec\phi_1}{b} \cdot y = -1, \tag{3}$$

$$\frac{-\tan\phi_1}{a} \cdot x + \frac{\sec\phi_1}{b} \cdot y = -1. \tag{4}$$

Making these equations simultaneous and combining by addition or subtraction, it is found that the vertices of the parallelogram lie in the asymptotes $y = \pm \frac{b}{a} x$.

Problems. — 1. Find the area of the ellipse whose axes are 8 and 6.

2. What are the eccentric angles of the vertices of the ellipse? of the ends of the focal ordinate to the transverse axis?

3. The extremity of a diameter of the ellipse $\frac{x^2}{a^2} + \frac{y^2}{b^2} = 1$ is (x_1, y_1), the extremity of the conjugate diameter (x_2, y_2). Find x_2 and y_2 in terms of x_1 and y_1.

4. Solve the same problem for the hyperbola.

5. In the hyperbola whose axes are 10 and 6 the length of a diameter is 15. Find the length of the conjugate diameter.

6. Find the lengths of the equal conjugate diameters of the ellipse whose axes are 12 and 8. Also the equation of this ellipse referred to its equal conjugate diameters.

CHAPTER X

SECOND DEGREE EQUATION

Art. 69.—Locus of Second Degree Equation

Write the general second degree equation in two variables in the form

$$ax^2 + 2bxy + cy^2 + 2dx + 2ey + f = 0. \qquad (1)$$

The problem is to determine the geometric figure represented by this equation when interpreted with respect to the rectangular axes X, Y. The equation of this geometric figure when referred to axes X_1, Y_1, parallel to X, Y, with origin at (x_0, y_0), becomes

$$ax_1^2 + 2bx_1y_1 + cy_1^2 + 2(ax_0 + by_0 + d)x_1 + 2(bx_0 + cy_0 + e)y_1$$
$$+ ax_0^2 + 2bx_0y_0 + cy_0^2 + 2dx_0 + 2ey_0 + f = 0. \qquad (2)$$

The geometric figure is symmetrical with respect to the new origin (x_0, y_0) if the coefficients of the terms in the first powers of the variables in equation (2) are zero. The coordinates of the center of symmetry of the figure are therefore determined by the equations $ax_0 + by_0 + d = 0$, $bx_0 + cy_0 + e = 0$. Whence $x_0 = \dfrac{eb - cd}{ac - b^2}$, $y_0 = \dfrac{db - ae}{ac - b^2}$. The center is a determinate finite point only when $ac - b^2 \neq 0$.

Suppose $ac - b^2 \neq 0$. The absolute term of equation (2) becomes

$$ax_0^2 + 2bx_0y_0 + cy_0^2 + 2dx_0 + 2ey_0 + f$$
$$= x_0(ax_0 + by_0 + d) + y_0(cy_0 + bx_0 + e) + dx_0 + ey_0 + f$$
$$= dx_0 + ey_0 + f = \frac{acf + 2bde - ae^2 - cd^2 - fb^2}{ac - b^2}.$$

Writing the last expression $\dfrac{\Delta}{ac-b^2}$, equation (2) becomes

$$ax_1^2 + 2bx_1y_1 + cy_1^2 + \dfrac{\Delta}{ac-b^2} = 0, \tag{3}$$

or
$$ax_1^2 + 2bx_1y_1 + cy_1^2 + k = 0, \tag{4}$$
where
$$k = dx_0 + ey_0 + f.$$

If $\Delta = 0$, equation (3) becomes

$$ax_1^2 + 2bx_1y_1 + cy_1^2 = 0, \tag{5}$$

which determines two values real or imaginary for $\dfrac{y_1}{x_1}$; that is, the equation resolves into two linear equations, and hence represents two straight lines. An equation which resolves into lower degree equations is called reducible, and the function of the coefficients, Δ, whose vanishing makes this resolution possible, is called the discriminant of the equation.

Turn the axes X_1, Y_1 about the origin (x_0, y_0) through an angle θ. Equation (4) becomes

$$(a\cos^2\theta + c\sin^2\theta + 2b\sin\theta\cos\theta)x_2^2$$
$$+ (a\sin^2\theta + c\cos^2\theta - 2b\sin\theta\cos\theta)y_2^2$$
$$+ 2\{(c-a)\sin\theta\cos\theta + b(\cos^2\theta - \sin^2\theta)\}x_2y_2 + k = 0.$$

Determine θ by equating to zero the coefficient of x_2y_2, whence $\tan 2\theta = \dfrac{2b}{a-c}$. Writing the resulting equation $Mx_2^2 + Ny_2^2 + k = 0$, it follows that

$$M + N = a + c, \quad M - N = (a-c)\cos(2\theta) + 2b\sin(2\theta).$$

From $\tan(2\theta) = \dfrac{2b}{a-c}$, $\sin(2\theta) = \dfrac{2b}{\{4b^2 + (a-c)^2\}^{\frac{1}{2}}}$,

$$\cos(2\theta) = \dfrac{a-c}{\{4b^2 + (a-c)^2\}^{\frac{1}{2}}}.$$

Therefore, $M + N = a + c$, $M - N = \{4b^2 + (a-c)^2\}^{\frac{1}{2}}$, and $MN = ac - b^2$. Now the equation $Mx_2^2 + Ny_2^2 + k = 0$ represents an ellipse referred to its axes when M and N have like signs, an hyperbola referred to its axes when M and N have

SECOND DEGREE EQUATION

unlike signs. Hence the second degree equation represents an ellipse when $ac - b^2 > 0$, an hyperbola when $ac - b^2 < 0$.

$\tan(2\theta) = \dfrac{2b}{a-c}$ determines two values for 2θ, and the radical $\{4b^2 + (a-c)^2\}^{\frac{1}{2}}$ has the double sign. To resolve the ambiguity take 2θ less than $180°$, which makes $\sin(2\theta)$ positive, and requires that the sign of the radical be the same as the sign of b. When $a - c$ and the radical have the same sign, $\cos(2\theta)$ is positive and 2θ is less than $90°$; when $a - c$ and the radical have different signs, $\cos(2\theta)$ is negative and 2θ is greater than $90°$.

The ambiguity may be resolved and the squares of the semi-axes calculated in this manner. The equation $\tan(2\theta) = \dfrac{2b}{a-c}$, written $\dfrac{2\tan\theta}{1-\tan^2\theta} = \dfrac{2b}{a-c}$, determines two values for $\tan\theta$. Call these values $\tan\theta_1$ and $\tan\theta_2$, and let θ_1 locate the X_2-axis, θ_2 the Y_2-axis. In the equation $ax_1^2 + 2bx_1y_1 + cy_1^2 + k = 0$, substitute $x_1 = r\cos\theta$, $y_1 = r\sin\theta$, and solve for r^2. There results $r^2 = -k \dfrac{1 + \tan^2\theta}{a + 2b\tan\theta + c\tan^2\theta}$. Calling the values of r^2 corresponding to $\tan\theta_1$ and $\tan\theta_2$ respectively r_1^2 and r_2^2, the equation of the ellipse or hyperbola referred to the axes X_2, Y_2, is $\dfrac{x_2^2}{r_1^2} + \dfrac{y_2^2}{r_2^2} = 1$.

When $ac - b^2 = 0$, the general equation becomes

$$ax^2 + 2a^{\frac{1}{2}}c^{\frac{1}{2}}xy + cy^2 + 2dx + 2ey + f = 0,$$

which may be written $(a^{\frac{1}{2}}x + c^{\frac{1}{2}}y)^2 + 2dx + 2ey + f = 0$. Transform to rectangular axes with $a^{\frac{1}{2}}x + c^{\frac{1}{2}}y = 0$ for X-axis, the origin unchanged. Then

$$\tan\theta = -\dfrac{a^{\frac{1}{2}}}{c^{\frac{1}{2}}} \quad \text{and} \quad \sin\theta = \dfrac{-a^{\frac{1}{2}}}{(a+c)^{\frac{1}{2}}}, \quad \cos\theta = \dfrac{c^{\frac{1}{2}}}{(a+c)^{\frac{1}{2}}}.$$

The transformation formulas become

$$x = \dfrac{c^{\frac{1}{2}}x_1 + a^{\frac{1}{2}}y_1}{(a+c)^{\frac{1}{2}}}, \quad y = \dfrac{-a^{\frac{1}{2}}x_1 + c^{\frac{1}{2}}y_1}{(a+c)^{\frac{1}{2}}}.$$

The transformed equation is

$$y_1^2 + 2\frac{a^{\frac{1}{2}}d + c^{\frac{1}{2}}e}{(a+c)^{\frac{1}{2}}}y_1 = 2\frac{a^{\frac{1}{2}}e - c^{\frac{1}{2}}d}{(a+c)^{\frac{1}{2}}}x_1 - \frac{f}{(a+c)^{\frac{1}{2}}},$$

which may be written in the form

$$(y_1 - n)^2 = 2\frac{a^{\frac{1}{2}}e - c^{\frac{1}{2}}d}{(a+c)^{\frac{1}{2}}}(x_1 - m),$$

the equation of a parabola whose parameter is $2\dfrac{a^{\frac{1}{2}}e - c^{\frac{1}{2}}d}{(a+c)^{\frac{1}{2}}}$, and whose vertex referred to the axes X_1, Y_1, is (m, n).

The condition $ac - b^2 = 0$ causes the center (x_0, y_0) of the conic section to go to infinity. Hence the parabola may be regarded as an ellipse or hyperbola with center at infinity. When the discriminant Δ also equals zero, the parabola becomes two straight lines intersecting at infinity; that is, two parallel straight lines.

It is now seen that every second degree equation in two variables interpreted in rectangular coordinates represents some variety of conic section.*

Problems.—Determine the variety, magnitude, and position of the conic sections represented by the following equations:

1. $14 x^2 - 4 xy + 11 y^2 - 44 x - 58 y + 71 = 0$.

$ac - b^2 = + 150$, therefore the equation represents an ellipse. The center is determined by the equations

$$14 x_0 - 2 y_0 - 22 = 0, \quad - 2 x_0 + 11 y_0 - 29 = 0,$$

---

* The three varieties of curves of the second order are plane sections of a right circular cone, which is for this reason called a cone of the second order. When the conic section becomes two parallel straight lines, the cone becomes a cylinder.

Newton (1642–1727) discovered that the curves of the third order are plane sections of five cones which have for bases the curves 21-25 on page 44. Plücker (1801-1868) showed that curves of the third order have 219 varieties.

SECOND DEGREE EQUATION

to be the point $(2, 3)$. $k = dx_0 + ey_0 + f = -60$. The axes are determined in direction by $\tan(2\theta) = -\frac{4}{3}$, whence $2\tan^2\theta - 3\tan\theta - 2 = 0$, $\tan\theta = 2$ or $-\frac{1}{2}$. $M + N = 25$, $MN = 150$. If the X-axis corresponds to $\tan\theta = 2$, $M - N$ must have the same sign as b. Therefore

$M - N = -5$, $M = 10$, $N = 15$.

The equation of the ellipse

$$10\,x_2^2 + 15\,y_2^2 = 60,$$

or

$$\frac{x_2^2}{6} + \frac{y_2^2}{4} = 1.$$

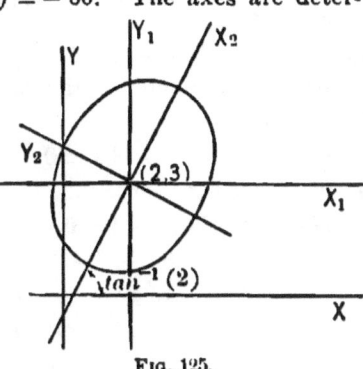

Fig. 125.

2. $x^2 - 3xy + y^2 + 10x - 10y + 21 = 0$.

$ac - b^2 = -\frac{5}{4}$, therefore the equation represents an hyperbola. The center, determined by the equations $x_0 - \frac{3}{2}y_0 + 5 = 0$, $-\frac{3}{2}x_0 + y_0 - 5 = 0$, is $(-2, 2)$. $k = dx_0 + ey_0 + f = +1$. The axes are determined in direction by $\tan(2\theta) = \infty$, whence $\theta_1 = 45°$, $\theta_2 = 135°$. By substituting in

$$r^2 = -k\frac{1 + \tan^2\theta}{a + 2b\tan\theta + c\tan^2\theta}, \quad r_1^2 = 2, \quad r_2^2 = -\frac{2}{5}.$$

The equation of the hyperbola referred to its own axes is $\frac{1}{2}x^2 - \frac{5}{2}y^2 = 1$.

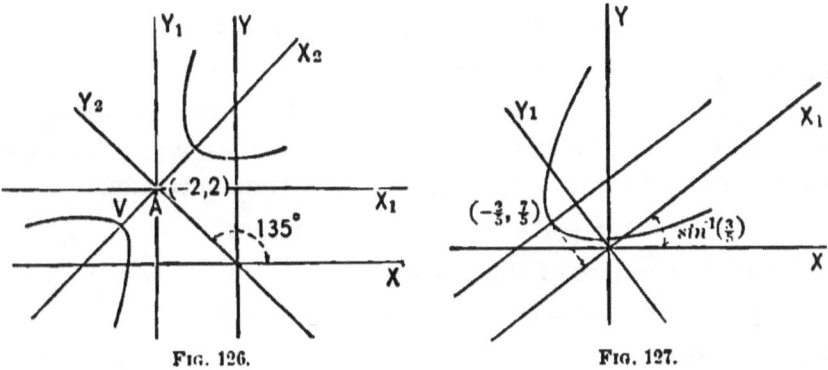

Fig. 126.   Fig. 127.

3. $9x^2 - 24xy + 16y^2 - 18x - 101y + 19 = 0$.

$ac - b^2 = 0$, therefore the equation represents a parabola. Write the equation in the form $(3x - 4y)^2 - 18x - 101y + 19 = 0$. Take $3x - 4y = 0$ as X-axis of a rectangular system of coordinates, the origin unchanged. Then $\tan\theta = \frac{3}{4}$, $\sin\theta = \frac{3}{5}$, $\cos\theta = \frac{4}{5}$, and the transformation formulas become $x = \dfrac{4x_1 - 3y_1}{5}$, $y = \dfrac{3x_1 + 4y_1}{5}$. The transformed equation is $25 y_1^2 - 75 x_1 - 70 y_0 + 19 = 0$, which may be written $(y_1 - \tfrac{7}{5})^2 = 3 (x_1 + \tfrac{2}{5})$. Hence the parameter of the parabola is 3, the vertex referred to the new axes $(-\tfrac{2}{5}, \tfrac{7}{5})$.

4. $y^2 + 2xy + 3x^2 - 4x = 0$.
5. $y^2 + 2xy - 3x^2 - 4x = 0$.
6. $y^2 - 2xy + x^2 + x = 0$.
7. $y^2 - 2xy + 2 = 0$.
8. $y^2 + 4xy + 4x^2 - 4 = 0$.
9. $3x^2 + 2xy + 3y^2 = 8$.
10. $4x^2 - 4xy + y^2 - 12x + 6y + 9 = 0$.
11. $x^2 - xy - 6y^2 = 6$.
12. $x^2 + xy + y^2 + x + y = 1$.
13. $3x^2 + 4xy + y^2 - 3x - 2y + 21 = 0$.
14. $5x^2 + 4xy + y^2 - 5x - 3y - 10 = 0$.
15. $4x^2 + 4xy + y^2 - 5x - 2y - 10 = 0$.

Art. 70. — Second Degree Equation in Oblique Coordinates

To determine the locus represented by
$$ax^2 + 2bxy + cy^2 + 2dx + 2cy + f = 0, \qquad (1)$$
when interpreted in oblique axes including an angle β, let
$$a'x'^2 + 2b'x'y' + c'y'^2 + 2d'x' + 2e'y' + f' = 0 \qquad (2)$$
be the result obtained by transforming the given equation to rectangular axes, the origin unchanged. Since (x, y) represents any point P referred to the oblique axes, and (x', y') the same point referred to rectangular axes, the expressions
$$x^2 + y^2 + 2xy \cos \beta \text{ and } x'^2 + y'^2$$
are each the square of the distance from P to the origin.

Hence $\qquad x^2 + y^2 + 2xy \cos \beta \equiv x'^2 + y'^2.$ $\qquad (3)$

By hypothesis
$$ax^2 + 2bxy + cy^2 \equiv a'x'^2 + 2b'x'y' + c'y'^2. \qquad (4)$$

Multiply the identity (3) by λ and add the product to (4). There results the identity
$$(a + \lambda) x^2 + 2 (b + \lambda \cos \beta) xy + (c + \lambda) y^2$$
$$\equiv (a' + \lambda) x'^2 + 2 b' x' y' + (c' + \lambda) y'^2.$$

Now any value of λ which makes the left-hand member of this identity a perfect square must also make the right-hand member a perfect square. The left-hand member is a perfect square when
$$\left(\frac{b+\lambda\cos\beta}{a+\lambda}\right)^2 = \frac{c+\lambda}{a+\lambda};$$
that is, when $\lambda^2 + \dfrac{a+c-2b\cos\beta}{\sin^2\beta}\lambda + \dfrac{ac-b^2}{\sin^2\beta} = 0.$

The right-hand member is a perfect square when
$$\lambda^2 + (a'+b')\lambda + a'c' - b'^2 = 0.$$
Since these equations determine the same values for λ,
$$a'c' - b'^2 = \frac{ac-b^2}{\sin^2\beta}.$$
Therefore $ac - b^2$ is greater than zero when $a'c' - b'^2$ is greater than zero. When $a'c' - b'^2 > 0$, equation (2) represents an ellipse when interpreted in rectangular coordinates. Consequently when $ac - b^2 > 0$ equation (1) represents an ellipse when interpreted in oblique coordinates. In like manner it follows that equation (1) interpreted in oblique coordinates represents an hyperbola when $ac - b^2 < 0$, a parabola when $ac - b^2 = 0$.

Problems. — 1. Two vertices of a triangle move along two intersecting straight lines. Find the curve traced by the third vertex.

From the figure are obtained the proportions of $\dfrac{y}{b} = \dfrac{\sin(\theta + a)}{\sin\omega},$

$$\frac{x}{a} = \frac{\sin(\theta + \omega - \beta)}{\sin\omega},$$

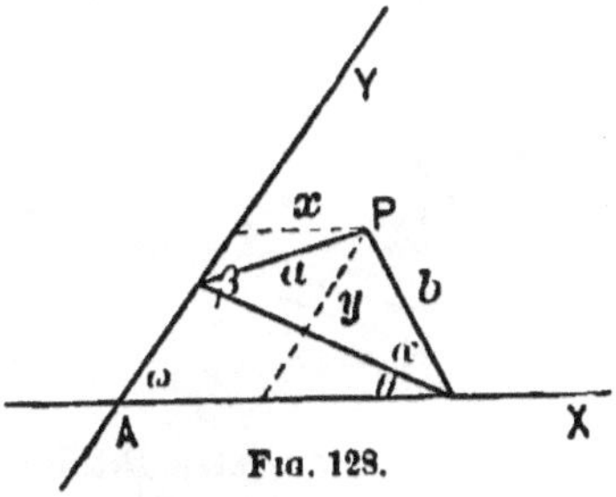

whence
$$\cos\theta = \frac{\dfrac{y}{b}\cos(\omega-\beta)\sin\omega - \dfrac{x}{a}\cos a\sin\omega}{\sin(a+\beta-\omega)},$$

$$\sin\theta = \frac{\dfrac{y}{b}\sin(\omega-\beta)\sin\omega - \dfrac{x}{a}\sin a\sin\omega}{\sin(a+\beta-\omega)}$$

FIG. 128.

Substituting in $\sin^2 \theta + \cos^2 \theta = 1$, there results

$$\frac{x^2}{a^2} - 2\frac{\sin(\alpha - \beta + \omega)}{ab} xy + \frac{y^2}{b^2} = \frac{\sin^2(\alpha + \beta - \omega)}{\sin^2 \omega},$$

the equation of an ellipse.

2. Find the envelope of a straight line which moves in such a manner that the sum of its intercepts on two intersecting straight lines is constant.

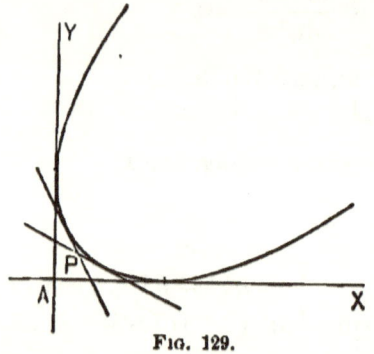

Fig. 129.

Let $\dfrac{x}{a} + \dfrac{y}{b} = 1$ be the moving straight line, then must $a + b = c$, where c is a constant. The equation of the straight line becomes $\dfrac{x}{a} + \dfrac{y}{c-a} = 1$, which may be written $a^2 + (y - x - c)a = cx$. The equation determines for every point $P(x, y)$ two values of a, to which correspond two lines of the system intersecting at (x, y). When these two values of a become equal, the point (x, y) becomes the intersection of consecutive positions of the line; that is, a point of the envelope of the line. Hence the point (x, y) of the envelope must satisfy the condition that the equation in a has equal roots. The equation of the envelope is therefore

$$(y - x - c)^2 + 4cx = 0,$$

which reduces to

$$y^2 - 2xy + x^2 - 2cy + 2cx + c^2 = 0$$

and represents a parabola.

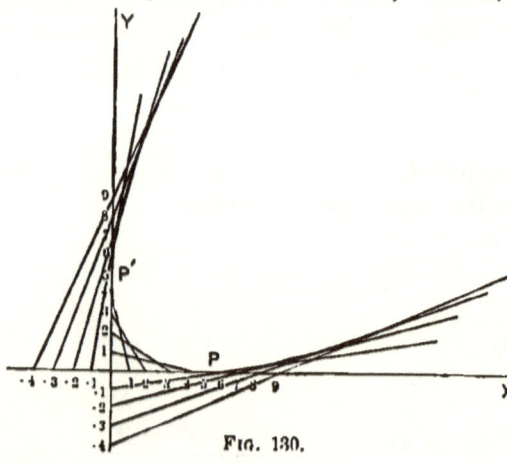

Fig. 130.

This problem furnishes a method frequently used to construct a parabola tangent to two given straight lines at points equidistant from their intersection. Mark on the lines starting at their intersection the equidistant points

1, 2, 3, 4, 5, 6, 7, 8, ···, −1, −2, −3, −4, −5, −6, −7, −8, ···.

If the given points are $+5$ on one line and $+5$ on the other, the straight lines joining the points of the given lines the sum of whose marks is $+5$ envelop the parabola required.

3. Through a fixed point a system of straight lines is drawn. Find the locus of the middle points of the segments of these lines included by the axes of reference.

4. Find the envelope of a straight line of constant length whose extremities slide in two fixed intersecting straight lines.

Art. 71.—Conic Section through Five Points

Let (x_1, y_1), (x_2, y_2), (x_3, y_3), (x_4, y_4) be four points of which no three are in the same straight line. Let $a = 0$ be the straight line through (x_1, y_1), (x_2, y_2); $b = 0$ the line through (x_2, y_2), (x_3, y_3); $c = 0$ the line through (x_3, y_3), (x_4, y_4); $d = 0$ the line through (x_4, y_4), (x_1, y_1). The equation $ac + kbd = 0$, where k is an arbitrary constant, represents a conic section through the four points. For, since a, b, c, d are linear, the equation $ac + kbd = 0$ is of the second degree, and must therefore represent a conic

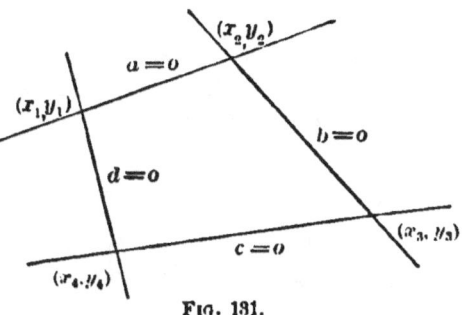

Fig. 131.

section. The equation is satisfied by $a = 0$ and $b = 0$, conditions which determine the point (x_2, y_2); by $a = 0$ and $d = 0$, determining the point (x_1, y_1); by $c = 0$, $b = 0$, determining (x_3, y_3); by $c = 0$, $d = 0$, determining (x_4, y_4). Since k is arbitrary, $ac + kbd = 0$ represents any one of an infinite number of conic sections through the four given points.

If the conic section is required to pass through a fifth point (x_5, y_5) not in the same straight line with any two of the four points (x_1, y_1), (x_2, y_2), (x_3, y_3), (x_4, y_4), the substitution of the coordinates of (x_5, y_5) in $ac + kbd = 0$ determines a single value for k. Therefore five points of which no three lie in the same straight line completely determine a conic section.

Problems. — Find the equations of conic sections through the five points.

1. $(1, 2)$, $(3, 5)$, $(-1, 4)$, $(-3, -1)$, $(-4, 3)$.

The equations of the sides of the quadrilateral whose vertices are the first four points are $a = 3x - 2y + 1 = 0$, $b = x - 4y + 17 = 0$, $c = 5x - 2y + 13 = 0$, $d = 3x - 4y + 5 = 0$. The equation of a conic section through these four points is therefore

$$(3x - 2y + 1)(5x - 2y + 13) + k(x - 4y + 17)(3x - 4y + 5) = 0.$$

Substituting the coordinates of the fifth point $(-4, 3)$, $k = \tfrac{221}{19}$. The equation of the conic section through the five points is

$$79 x^2 - 320 xy + 301 y^2 + 1101 x - 1065 y + 1586 = 0.$$

2. $(2, 3)$, $(0, 4)$, $(-1, 5)$, $(-2, -1)$, $(1, -2)$.
3. $(1, 3)$, $(4, -6)$, $(0, 0)$, $(9, -9)$, $(16, 12)$.
4. $(-4, -2)$, $(2, 1)$, $(-6, 3)$, $(0, 0)$, $(2, -1)$.
5. $(-\tfrac{2}{3}, -\tfrac{1}{3})$, $(2, 1)$, $(\tfrac{5}{3}, 2)$, $(-\tfrac{2}{3}, -3)$, $(\tfrac{5}{3}, -\tfrac{2}{3})$.
6. $(3, \sqrt{5})$, $(-2, 0)$, $(-4, -\sqrt{12})$, $(3, -\sqrt{5})$ $(2, 0)$.
7. $(1, 2)$, $(2, 1)$, $(3, -2)$, $(0, 4)$, $(3, 0)$.
8. $(2, 3)$, $(-2, 3)$, $(4, 1)$, $(1, 3)$, $(0, 0)$.

Art. 72. — Conic Sections Tangent to Given Lines

Let $a = 0$ and $b = 0$ represent two straight lines intersected by the straight line $c = 0$. The equation $ab - kc^2 = 0$ represents a conic section tangent to the lines $a = 0$, $b = 0$ at the points of intersection of $c = 0$. For the equation $ab - kc^2 = 0$ is of the second degree, and the points of intersection of the line $a = 0$ with $ab - kc^2 = 0$ coincide at the point of intersection of the lines $a = 0$, $c = 0$, which makes $a = 0$ tangent to the conic section. For a like reason $b = 0$ is tangent to

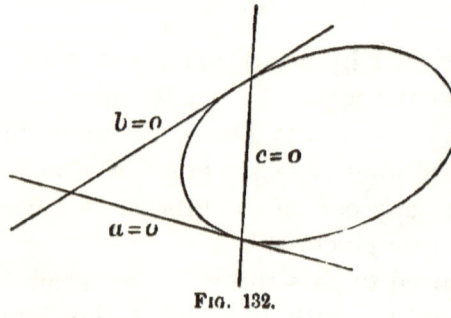

Fig. 132.

SECOND DEGREE EQUATION

$ab - kc^2 = 0$. Since k is arbitrary, an infinite number of conic sections can be drawn tangent to the given lines at the given points.

The equation of a conic section tangent to the lines $x = 0$, $y = 0$ at the points $(a, 0)$, $(0, b)$ is

$$\left(\frac{x}{a} + \frac{y}{b} - 1\right)^2 - Kxy = 0. \quad (1)$$

The points of intersection of this conic section and the line MN, $\frac{x}{m} + \frac{y}{n} = 1$, lie in the locus of the equation

$$\left(\frac{x}{a} + \frac{y}{b} - \frac{x}{m} - \frac{y}{n}\right)^2 = Kxy. \quad (2)$$

This last equation is homogeneous of the second degree, and hence represents two straight lines from the origin through the points of intersection of $\frac{x}{m} + \frac{y}{n} - 1 = 0$

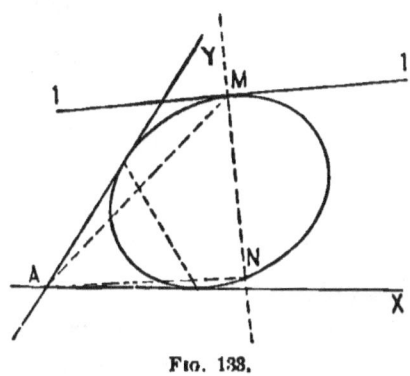

Fig. 133.

and $\left(\frac{x}{a} + \frac{y}{b} - 1\right)^2 - Kxy = 0$. The straight lines represented by equation (2) coincide, and $\frac{x}{m} + \frac{y}{n} - 1 = 0$ is tangent to

$$\left(\frac{x}{a} + \frac{y}{b} - 1\right)^2 - Kxy = 0,$$

when

$$\left(\frac{x}{a} + \frac{y}{b} - \frac{x}{m} - \frac{y}{n}\right)^2 - Kxy$$

is a perfect square; that is, when

$$\left(\frac{1}{a} - \frac{1}{m}\right)^2 \left(\frac{1}{b} - \frac{1}{n}\right)^2 = \left\{\left(\frac{1}{a} - \frac{1}{m}\right)\left(\frac{1}{b} - \frac{1}{n}\right) - \frac{K}{2}\right\}^2,$$

whence
$$K = 4\left(\frac{1}{a} - \frac{1}{m}\right)^2\left(\frac{1}{b} - \frac{1}{n}\right). \tag{3}$$

Similarly, $\dfrac{x}{m} + \dfrac{y}{n} - 1 = 0$ is tangent to (1) when

$$K = 4\left(\frac{1}{a} - \frac{1}{m_1}\right)\left(\frac{1}{b} - \frac{1}{n_1}\right). \tag{4}$$

Equations (3) and (4) determine the values of $\dfrac{1}{a}$ and $\dfrac{1}{b}$ in terms of the arbitrary constant K, which shows that an infinite number of conic sections can be drawn tangent to four straight lines no three of which pass through a common point. If $\dfrac{x}{m_2} + \dfrac{y}{n_2} = 1$ is also tangent to the conic section represented by equation (1),

$$K = 4\left(\frac{1}{a} - \frac{1}{m_2}\right)\left(\frac{1}{b} - \frac{1}{n_2}\right). \tag{5}$$

Equations (3), (4), (5) determine $\dfrac{1}{a}, \dfrac{1}{b}$, and K uniquely, proving that only one conic section can be drawn tangent to five straight lines no three of which pass through a common point. This proposition is the reciprocal of the proposition of Art. 71 and might have been demonstrated by the method of reciprocal polars.

Problems.—1. Find the equation of the parabola tangent to two straight lines including an angle of 60° at points whose distances from their point of intersection are 2 and 4.

2. Find the equation of the conic section tangent to two straight lines including an angle of 45° at (3, 0), (5, 0), and containing the point (7, 8), the given straight lines being the axes of reference.

Art. 73.—Similar Conic Sections

The points $P(x, y)$ and $P_1(mx, my)$ lie in the same straight line through the origin O, and $OP_1 = m \cdot OP$. The distance between any two positions of P_1, (mx', my'), (mx'', my'') is m

times the distance between the corresponding positions of P, (x', y'), (x'', y''). For

$$\{(mx' - mx'')^2 + (my' - my'')^2\}^{\frac{1}{2}} = m\{(x' - x'')^2 + (y' - y'')^2\}^{\frac{1}{2}}.$$

Representing the point P by (x, y), the point P_1 by (X, Y), when (x, y) traces a geometric figure, the point (X, Y) traces a figure to scale m times as large. The effect of the substitution $x = \dfrac{X}{m}$, $y = \dfrac{Y}{m}$ is therefore simply to change the scale of the drawing. Figures thus related are said to be similar. When the two equations $f(x, y) = 0$, $f\left(\dfrac{X}{m}, \dfrac{Y}{m}\right) = 0$ are interpreted in the same axes, their loci are similar and similarly placed; when interpreted in different axes but including the same angle, the loci are similar. Ellipses similar to $\dfrac{x^2}{a^2} + \dfrac{y^2}{b^2} = 1$ are represented by $\dfrac{X^2}{m^2 a^2} + \dfrac{Y^2}{m^2 b^2} = 1$. All ellipses of a similar system have the same eccentricity, for

$$e^2 = \frac{m^2 a^2 - m^2 b^2}{m^2 a^2}$$
$$= \frac{a^2 - b^2}{a^2}.$$

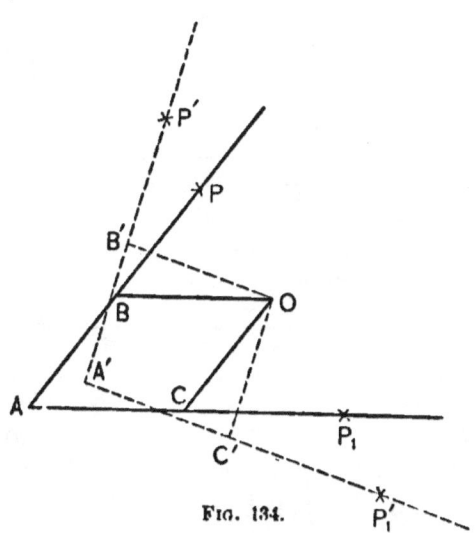

Fig. 134.

In like manner, all hyperbolas of a similar system $\dfrac{X^2}{m^2 a^2} - \dfrac{Y^2}{m^2 b^2} = 1$ have the same eccentricity. The parabolas similar to $y^2 = 2px$ are represented by the equation $Y^2 = 2pmX$.

Taking as corresponding points

$$P(x, y), \quad P_1(mx, my), \quad P_2(-mx, -my),$$

L

the figure traced by P_1 is similar to that traced by P, the figure traced by P_2 is symmetrical to that traced by P_1.

The change of scale of a drawing may be effected mechanically by means of an instrument called the pantograph, which consists of four rods jointed together in such a manner as to form a parallelogram $ABOC$ with sides of constant length, but whose angles may be changed with perfect freedom. On the rods AB and AC fix two points P and P_1 in a straight line with O. If the point O is fixed in the plane, and the point P is made to take any new position P'', and the corresponding position of P_1 is P_1', the points P', O, P_1' in Fig. 134 are always in a straight line, the triangles $P_1'C'O$ and $P_1'A'P''$ are similar and hence

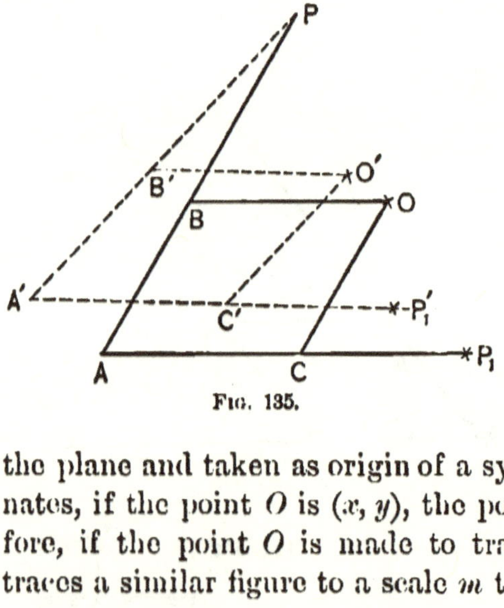

$$\frac{OP_1'}{OP''} = \frac{P_1'C'}{C'A'} = \frac{P_1C}{CA},$$

a constant which may be denoted by m. Taking O as origin of a system of rectangular coordinates, if P is (x, y), P_1 is

$$(-mx, -my).$$

Fig. 135.

If the point P is fixed in the plane and taken as origin of a system of rectangular coordinates, if the point O is (x, y), the point P_1 is (mx, my). Therefore, if the point O is made to trace any locus, the point P_1 traces a similar figure to a scale m times as large.

Art. 74.—Confocal Conic Sections

The equation
$$\frac{x^2}{a^2 + \lambda} + \frac{y^2}{b^2 + \lambda} = 1, \tag{1}$$

where $a^2 > b^2$ represents an ellipse when $\lambda > -b^2$, an hyperbola when $-a^2 < \lambda < -b^2$, an imaginary locus when $\lambda < -a^2$. The

SECOND DEGREE EQUATION

distance from focus to center of the ellipses and hyperbolas represented by equation (1) is $\{a^2 + \lambda - b^2 - \lambda\}^{\frac{1}{2}} = (a^2 - b^2)^{\frac{1}{2}}$. Hence equation (1) when interpreted for different values of λ in the same rectangular axes represents ellipses and hyperbolas having common foci; that is, a system of confocal conic sections.

Through every point (x', y') of the plane there passes one ellipse and one hyperbola of the confocal system

$$\frac{x^2}{a^2 + \lambda} + \frac{y^2}{b^2 + \lambda} = 1.$$

For the conic sections passing through (x', y') correspond to the values of λ satisfying the equation

$$\frac{x'^2}{a^2 + \lambda} + \frac{y'^2}{b^2 + \lambda} = 1. \quad (2)$$

This function of λ,

$$\frac{x'^2}{a^2 + \lambda} + \frac{y'^2}{b^2 + \lambda} - 1,$$

is negative when $\lambda = +\infty$, positive just before λ becomes $-b^2$, negative when λ is just less than $-b^2$ and again positive when λ is just greater than $-a^2$. Hence equation (2) determines for λ two values, one between $+\infty$ and $-b^2$, the other between $-b^2$ and $-a^2$.

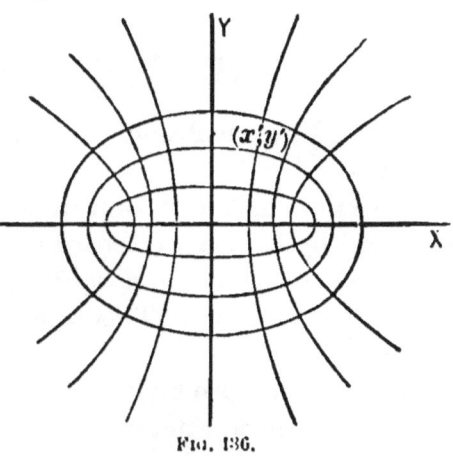

Fig. 136.

The ellipse and hyperbola of the confocal system

$$\frac{x^2}{a^2 + \lambda} + \frac{y^2}{b^2 + \lambda} = 1$$

through the point (x', y') intersect at right angles.

Let λ_1 and λ_2 be the values of λ satisfying the equation $\frac{x'^2}{a^2 + \lambda} + \frac{y'^2}{b^2 + \lambda} = 1.$ Then

$$\frac{x^2}{a^2 + \lambda_1} + \frac{y^2}{b^2 + \lambda_1} = 1, \quad \frac{x^2}{a^2 + \lambda_2} + \frac{y^2}{b^2 + \lambda_2} = 1$$

represent ellipse and hyperbola through (x', y'). The tangents to this ellipse and hyperbola at (x', y') are

$$\frac{xx'}{a^2+\lambda_1} + \frac{yy'}{b^2+\lambda_1} = 1, \tag{1}$$

$$\frac{xx'}{a^2+\lambda_2} + \frac{yy'}{b^2+\lambda_2} = 1. \tag{2}$$

From the equations

$$\frac{x'^2}{a^2+\lambda_1} + \frac{y'^2}{b^2+\lambda_1} = 1, \quad \frac{x'^2}{a^2+\lambda_2} + \frac{y'^2}{b^2+\lambda_2} = 1,$$

is obtained by subtraction

$$\frac{x'^2}{(a^2+\lambda_1)(a^2+\lambda_2)} + \frac{y'^2}{(b^2+\lambda_1)(b^2+\lambda_2)} = 0,$$

which is the condition of perpendicularity of tangents (1) and (2).

Since through every point in the plane there passes one ellipse and one hyperbola of the confocal system, the point of the plane is determined by specifying the ellipse and hyperbola in which the point lies. This leads to a system of elliptic coordinates.

If heat flows into an infinite plane disc along a finite straight line at a uniform rate, when the heat conditions have become permanent, the isothermal lines are the ellipses, the lines of flow of heat the hyperbolas of the confocal system. The same is true if instead of heat any fluid flows over the disc, or if an electric or magnetic disturbance enters along the straight line.

CHAPTER XI

LINE COORDINATES

Art. 75. — Coordinates of a Straight Line

If the equation of a straight line is written in the form $ux + vy + 1 = 0$, u and v are the negative reciprocals of the intercepts of the line on the axes. To every pair of values of u and v there corresponds one straight line, and conversely; that is, there is a "one-to-one correspondence" between the symbol (u, v) and the straight lines of the plane. u and v are called line coordinates.*

If (u, v) is fixed, the equation $ux + vy + 1 = 0$ expresses the condition that the point (x, y) lies in the straight line (u, v). The system of points on a straight line is called a range of points. Hence a first degree point equation represents a range of points and determines a straight line.

If (x, y) is fixed, $ux + vy + 1 = 0$ expresses the condition that the line (u, v) passes through the point (x, y). The system of lines through a point is called a pencil of rays. Hence a first degree line equation represents a pencil of rays and determines a point.

The equations $ux_1 + vy_1 + 1 = 0$, $ux_2 + vy_2 + 1 = 0$ determine the points (x_1, y_1), (x_2, y_2) respectively. $\left(\dfrac{x_1 + \lambda x_2}{1 + \lambda}, \dfrac{y_1 + \lambda y_2}{1 + \lambda} \right)$ represents for each value of λ one point of the line through (x_1, y_1), (x_2, y_2). λ is the ratio of the segments into which the point corresponding to λ divides the finite line from (x_1, y_1) to

---

* Plücker in Germany and Chasles in France developed the use of line coordinates at about the same time (1829).

(x_2, y_2). There is a "one-to-one correspondence" between λ and the points of the line through (x_1, y_1), (x_2, y_2).

$$u\frac{x_1 + \lambda x_2}{1 + \lambda} + v\frac{y_1 + \lambda y_2}{1 + \lambda} + 1 = 0,$$

which reduces to $(ux_1 + vy_1 + 1) + \lambda(ux_2 + vy_2 + 1) = 0$, is the line equation of the point λ. Denoting $ux_1 + vy_1 + 1$ by L, $ux_2 + vy_2 + 1$ by M, $L + \lambda M = 0$ represents the range of points determined by $L = 0$, $M = 0$.

The rays of the pencil determined by the lines

$$u_1 x + v_1 y + 1 = 0, \ u_2 x + v_2 y + 1 = 0$$

are represented by the equation

$$(u_1 x + v_1 y + 1) + \lambda(u_2 x + v_2 y + 1) = 0,$$

which may be written $\dfrac{u_1 + \lambda u_2}{1 + \lambda} x + \dfrac{v_1 + \lambda v_2}{1 + \lambda} + 1 = 0.$

Hence $\left(\dfrac{u_1 + \lambda u_2}{1 + \lambda}, \ \dfrac{v_1 + \lambda v_2}{1 + \lambda} \right)$

are the lines of the pencil determined by (u_1, v_1), (u_2, v_2). There is a "one-to-one correspondence" between λ and the rays of the pencil. Denoting $u_1 x + v_1 y + 1$ by P, $u_2 x + v_2 y + 1$ by Q, $P + \lambda Q = 0$ represents the rays of the pencil determined by $P = 0$, $Q = 0$.

Problems. — 1. Construct the lines $(4, 1)$; $(-2, 5)$; $(-\frac{1}{3}, -\frac{1}{4})$.

2. Construct the pencil represented by $3u - 2v + 1 = 0$.

3. Construct the range represented by $2x - 3y + 1 = 0$.

4. Locate the point determined by $4u + 5v + 1 = 0$.

5. Draw the line determined by $3x - 5y + 1 = 0$.

6. Write the equation of the range of points determined by
$$2u - 3v + 1 = 0, \ \tfrac{1}{3}u + \tfrac{1}{2}v + 1 = 0.$$

7. Write the equation of the pencil of rays determined by
$$2x - 3y + 1 = 0, \ \tfrac{1}{3}x + \tfrac{1}{2}y + 1 = 0.$$

Art. 76. — Line Equations of the Conic Sections

The equation of the tangent to the ellipse $\frac{x^2}{a^2} + \frac{y^2}{b^2} = 1$ at (x_0, y_0) is $\frac{xx_0}{a^2} + \frac{yy_0}{b^2} = 1$. Comparing the equation of the tangent with $ux + vy + 1 = 0$ it is seen that the line coordinates of the tangent are $u = -\frac{x_0}{a^2}$, $v = -\frac{y_0}{b^2}$, whence $x_0 = -a^2 u$, $y_0 = -b^2 v$. If the point of tangency (x_0, y_0) generates the ellipse $\frac{x^2}{a^2} + \frac{y^2}{b^2} = 1$, the tangent (u, v) envelopes the ellipse. Hence the line equation of the ellipse, when the reference axes are the axes of the ellipse, is $a^2 u^2 + b^2 v^2 = 1$.

Problems. — 1. Show that the line equation of the circle $x^2 + y^2 = r^2$ is $u^2 + v^2 = \frac{1}{r^2}$.

2. Show that the line equation of the hyperbola
$$\frac{x^2}{a^2} - \frac{y^2}{b^2} = 1 \text{ is } a^2 u^2 - b^2 v^2 = 1.$$

3. Show that the line equation of the parabola $y^2 = 2px$ is $pv^2 = 2u$. Construct the envelopes of the equations

4. $\frac{1}{u} + \frac{1}{v} = -5$. 6. $u^2 + v^2 = \frac{1}{9}$. 8. $9u^2 - 4v^2 = 1$.

5. $uv = \frac{1}{8}$. 7. $9u^2 + 4v^2 = 1$. 9. $8v^2 - u = 0$.

Art. 77. — Cross-ratio of Four Points

The double ratio $\frac{CA}{CB} \div \frac{DA}{DB}$ is called the cross-ratio of the four points A, B, C, D, and is denoted by the symbol $(ABCD)$. If the point A is denoted by $L = 0$, the point B by $M = 0$, the points C and D respectively by $L + \lambda_1 M = 0$ and $L + \lambda_2 M = 0$, it follows that $\frac{CA}{CB} = \lambda_1$, $\frac{DA}{DB} = \lambda_2$, and $(ABCD) = \frac{\lambda_1}{\lambda_2}$. Take any four points of

Fig. 137.

the range $L + \lambda M = 0$ corresponding to $\lambda_1, \lambda_2, \lambda_3, \lambda_4$, and represent $L + \lambda_1 M$ by L_1, $L + L_2 M$ by M_1, whence $L + \lambda_3 M$ is represented by $L_1 - \dfrac{\lambda_1 - \lambda_3}{\lambda_2 - \lambda_3} M_1$, $L + \lambda_4 M$ by $L_1 - \dfrac{\lambda_1 - \lambda_4}{\lambda_2 - \lambda_4} M_1$. The four points corresponding to $\lambda_1, \lambda_2, \lambda_3, \lambda_4$ are represented by the equations

$$L_1 = 0, \quad M_1 = 0, \quad L_1 - \dfrac{\lambda_1 - \lambda_3}{\lambda_2 - \lambda_3} M_1 = 0, \quad L_1 - \dfrac{\lambda_1 - \lambda_4}{\lambda_2 - \lambda_4} M_1 = 0,$$

and their cross-ratio is $\dfrac{\lambda_1 - \lambda_3}{\lambda_2 - \lambda_3} \dfrac{\lambda_2 - \lambda_4}{\lambda_1 - \lambda_4}$. Since the four points $\lambda_1, \lambda_2, \lambda_3, \lambda_4$ can be arranged in 24 different ways, the cross-ratio of four points takes 24 different forms, but these 24 different forms are seen to have only six different values.

$L + \lambda M = 0$, $L' + \lambda M' = 0$ represent two ranges of points. By making the point of one range determined by a value of λ correspond to the point of the other range determined by the same value of λ, a "one-to-one correspondence" is established between the points of the two ranges, and the cross-ratio of any four points of one range equals the cross-ratio of the corresponding four points of the second range. Such ranges are called projective.

Art. 78.—Second Degree Line Equations

Remembering that each of the equations

$$L + \lambda M = 0, \qquad L' + \lambda M' = 0$$

for any value of λ represents the entire pencil of rays through the point of the range corresponding to λ, it is evident that the equation $LM' - L'M = 0$, obtained by eliminating λ between $L + \lambda M = 0$, $L' + \lambda M' = 0$, represents the system of lines joining the corresponding points of the two projective point ranges. This equation is a second degree line equation, and it becomes necessary to determine the locus enveloped by the lines represented by the equation.

Let $ux + vy + 1 = 0$ represent any point (x, y) of the plane. Writing the values of L, M, L', M' in full, the elimination of u and v from the equations $ux + vy + 1 = 0$,

$$ux_1 + vy_1 + 1 + \lambda(ux_2 + vy_2 + 1) = 0,$$
$$ux' + vy' + 1 + \lambda(ux'' + vy'' + 1) = 0,$$

determines a quadratic equation in λ with coefficients of the first degree in (x, y), $G\lambda^2 + H\lambda + K = 0$. To the two values of λ which satisfy this equation there correspond the tangents from (x, y) to the envelope of $LM' - L'M = 0$. When these tangents coincide, the point (x, y) lies on the envelope.

$$4H^2 - GK = 0$$

causes the coincidence of the tangents, and is therefore the point equation of the envelope. The point equation being of the second degree, the envelope is a conic section.

The degree of a line equation denotes the number of tangents that can be drawn from any point in the plane to the curve represented by the equation, and is called the class of the curve.

ART. 79.—Cross-ratio of a Pencil of Four Rays

Let a pencil of four rays,

$$P = 0, \quad Q = 0, \quad P + \lambda_1 Q = 0, \quad P + \lambda_2 Q = 0,$$

be cut by any transversal in the four points A, B, C, D. p is the common altitude of the triangles whose common vertex is O, and whose bases lie in the transversal. Then

$p \cdot CA = OA \cdot OC \cdot \sin COA,$ $\quad p \cdot DA = OA \cdot OD \cdot \sin DOA,$
$p \cdot CB = OC \cdot OB \cdot \sin COB,$ $\quad p \cdot DB = OD \cdot OB \cdot \sin DOB,$

and $(ABCD) = \dfrac{\sin COA}{\sin COB} \div \dfrac{\sin DOA}{\sin DOB}$. This double sine ratio is called the cross-ratio of the pencil of four rays. It is evident that central projection does not alter the cross-ratio of four points in a straight line.

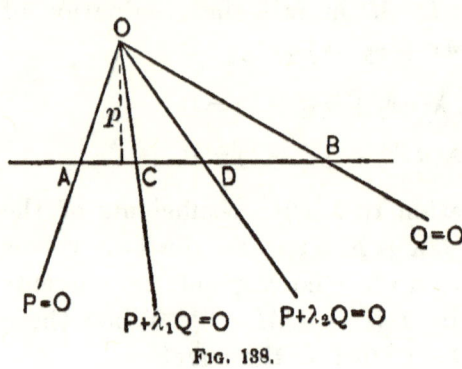

Fig. 139.

Writing the equation $P + \lambda Q = 0$ in the complete form

$$u_1 x + v_1 y + 1 + \lambda (u_2 x + v_2 y + 1) = 0,$$

and this in the form

$$\frac{u_1 x + v_1 y + 1}{(u_1^2 + v_1^2)^{\frac{1}{2}}} + \lambda \frac{(u_2^2 + v_2^2)^{\frac{1}{2}}}{(u_1^2 + v_1^2)^{\frac{1}{2}}} \frac{u_2 x + v_2 y + 1}{(u_2^2 + v_2^2)^{\frac{1}{2}}} = 0,$$

the factor $\lambda \dfrac{(u_2^2 + v_2^2)^{\frac{1}{2}}}{(u_1^2 + v_1^2)^{\frac{1}{2}}}$ is seen to be the negative ratio of the distances from any point of the line $P + \lambda Q = 0$ to the lines $P = 0$, $Q = 0$. Hence

$$\frac{Ca}{Cb} = \frac{\sin COA}{\sin COB} = -\lambda_1, \qquad \frac{Dd'}{Db'} = \frac{\sin DOA}{\sin DOB} = -\lambda_2,$$

and $\dfrac{\sin COA}{\sin COB} \div \dfrac{\sin DOA}{\sin DOB} = \dfrac{\lambda_1}{\lambda_2} =$ the cross-ratio of the four rays

$$P = 0, \quad Q = 0, \quad P + \lambda_1 Q = 0, \quad P + \lambda_2 Q = 0.$$

Representing

$P + \lambda_1 Q$ by P_1, $P + \lambda_2 Q$ by Q_1, $P + \lambda_3 Q$

is represented by

$$P_1 - \frac{\lambda_1 - \lambda_3}{\lambda_2 - \lambda_3} Q_1, \quad P + \lambda_4 Q \text{ by } P_1 - \frac{\lambda_1 - \lambda_4}{\lambda_2 - \lambda_4} Q_1.$$

Hence the cross-ratio of the four points of the pencil $P + \lambda Q = 0$ corresponding to λ_1, λ_2, λ_3, λ_4 is $\dfrac{\lambda_1 - \lambda_3}{\lambda_2 - \lambda_3} \dfrac{\lambda_2 - \lambda_4}{\lambda_1 - \lambda_4}$.

By making the ray of $P+\lambda Q=0$ determined by a value of λ correspond to the ray of $P'+\lambda Q'=0$ determined by the same value of λ, a "one-to-one correspondence" is established between the rays of the two pencils, and the cross-ratio of any four rays of one pencil equals the cross-ratio of the corresponding four rays of the other pencil. Such pencils are called projective pencils.

The equation of the locus of the points of intersection of the corresponding rays of the two projective pencils $P+\lambda Q=0$, $P'+\lambda Q'=0$ is $PQ'-P'Q=0$. This is a second degree point equation and represents a conic section.*

Art. 80. — Construction of Projective Ranges and Pencils

If there exists a "one-to-one correspondence" between the points of two ranges, between the rays of two pencils, or between the points of a range and the rays of a pencil, the ranges and pencils are projective.

Let $P=0$, $Q=0$, determining the range or pencil $P+\lambda Q=0$, correspond to $P_1=0$, $Q_1=0$, determining the range or pencil $P_1+\lambda_1 Q_1=0$, and let a "one-to-one correspondence" exist between the elements λ of the first system and the elements λ_1 of the second system. This "one-to-one correspondence" interpreted algebraically means that λ_1 is a linear function of λ; that is, $\lambda_1 = \dfrac{a\lambda + b}{c\lambda + d}$. By hypothesis, $\lambda_1 = 0$ when $\lambda = 0$, and $\lambda_1 = \infty$ when $\lambda = \infty$, hence $b=0$, $c=0$, and $\lambda_1 = \dfrac{a}{d}\lambda$. Let $\lambda = l$ and $\lambda_1 = l_1$ be a third pair of corresponding ele-

---

* A complete projective treatment of conic sections is developed in Steiner's Theorie der Kegelschnitte, 1866, and in Chasles' Géométrie Supérieure, 1852, and in Cremona's Elements of Projective Geometry, translated from the Italian.

ments; then $\frac{a}{d} = \frac{l_1}{l}$, $\lambda_1 = \frac{l_1}{l}\lambda$, and the equations of the systems $P + \lambda Q = 0$, $P_1 + \lambda_1 Q_1 = 0$ become $P + \lambda Q = 0$, $lP_1 + \lambda l_1 Q_1 = 0$.

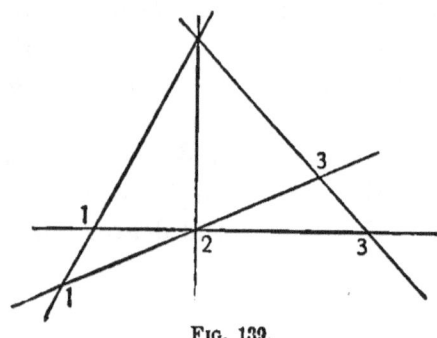

Fig. 139.

Now the elements of
$$lP_1 + \lambda l Q_1 = 0$$
are the elements of
$$P_1 + \lambda Q_1 = 0,$$
hence the systems between whose elements there exists a "one-to-one correspondence" are the projective systems
$$P + \lambda Q = 0, \quad P_1 + \lambda Q_1 = 0.$$

This analysis also shows that the correspondence of three elements of one system to three elements of another makes the systems projective.

Projective systems are constructed geometrically, as follows: Let the points 1, 2, 3 on one straight line mm correspond to

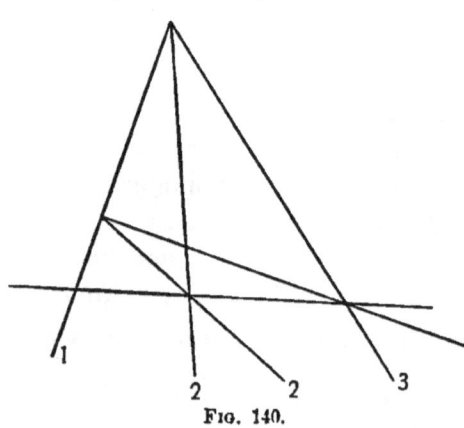

Fig. 140.

the points 1, 2, 3, respectively, on another straight line nn. Place the two lines with one pair of corresponding points 2, 2 in coincidence. Join the point of intersection 0 of the lines through 1, 1 and 3, 3 with 2. Take the points of intersection of lines through 0 with mm and nn as corresponding

points, and a "one-to-one correspondence" is established between the points of the ranges mm, nn, which are therefore projective.

In like manner, if three rays 1, 2, 3 of pencil m correspond to the rays 1, 2, 3, respectively, of pencil n, by placing the corresponding rays 1, 1 in coincidence, and drawing the line 00 through the points of intersection of the corresponding rays 2, 2 and 3, 3, and taking rays from m and n to any point of 00 as corresponding rays, a "one-to-one correspondence" is established between the rays of the two pencils, and the pencils are projective.

Art. 81.— Conic Section through Five Points

It is now possible by the aid of the ruler only to construct a conic section through five points or tangent to five lines. Take two of the given points 1, 2 as the vertices of pencils, the pairs of lines from 1 and 2 to the remaining three points 3, 4, 5, respectively, as corresponding rays of projective pencils. The

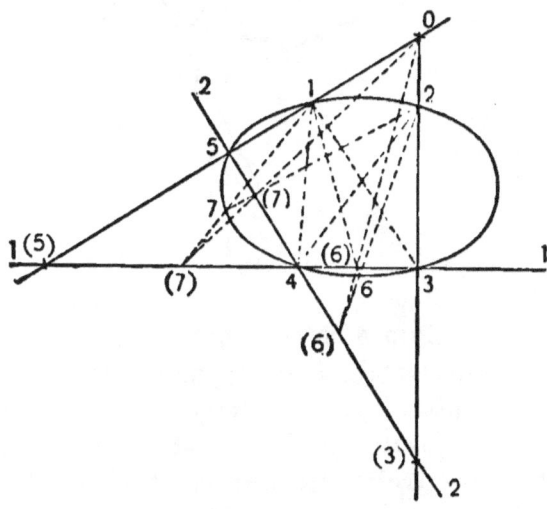

Fig. 141.

line 11 is a transversal of pencil 1, 22 of pencil 2. 0, the intersection of 51 and 32, is the vertex of a pencil of which 11 and 22 are transversals. Hence the pencils 1 and 2 are projective, and corresponding rays are rays to the points of intersection of the rays of pencil 0 with 11 and 22. The intersections of these corresponding rays are points of the required conic section.

Take two of the five given lines 11 and 22 as bearers of point ranges on which the points of intersection of the other lines

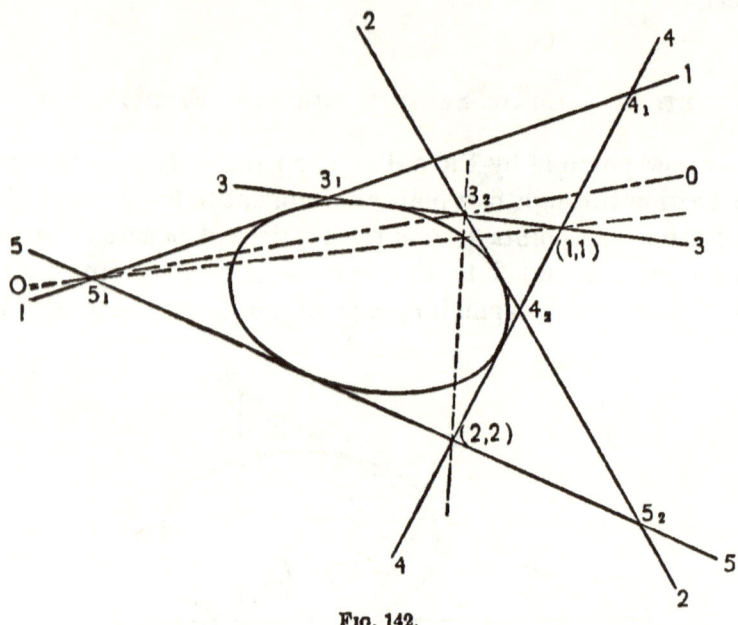

Fig. 142.

33, 44, 55 respectively are corresponding points. The line 00 is a common transversal of the pencils (11), (22). Hence corresponding points of the projective ranges 11 and 22 are located by the intersection with 11 and 22 of lines connecting (11) and (22) respectively with any point of 00. The straight lines connecting corresponding points are tangents to the required conic section.

Notice that the construction of the conic section tangent to five straight lines is the exact reciprocal of the construction of the conic section through five points.

The figure formed by joining by straight lines six arbitrary points on a conic section in any order whatever is called a six-point. Taking 1 and 5 as vertices of pencils whose corresponding rays are determined by the points 2, 3, 4, the points of intersection of 16 with 11 and of 56 with 55 must lie in the same ray of the auxiliary pencil O; that is, in any six-point of a conic section the intersection of the three pairs of opposite sides are in a straight line. This is Pascal's theorem.*

Reciprocating Pascal's theorem, Brianchon's theorem is obtained. — In the figure formed by drawing tangents to a conic section at six arbitrary points in any order whatever (a six-side of a conic section), the straight lines joining the three pairs of opposite vertices pass through a common point.†

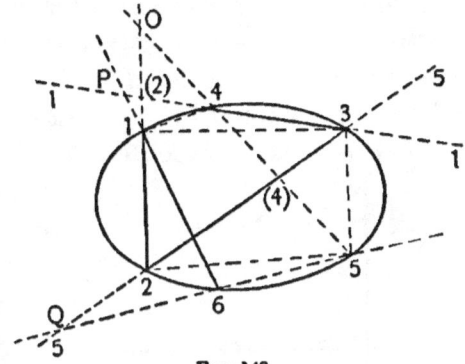

FIG. 143.

By Pascal's theorem any number of points on a conic section through five points may be located by the aid of the ruler; by Brianchon's theorem any number of tangents to a conic section tangent to five straight lines may be drawn by the aid of the ruler.

* Discovered by Pascal, 1640.
† Discovered by Brianchon, 1806.

CHAPTER XII

ANALYTIC GEOMETRY OF THE COMPLEX VARIABLE

Art. 82.—Graphic Representation of the Complex Variable

The expression $x + iy$, where x and y are real variables and i stands for $\sqrt{-1}$, is called the complex variable, and is frequently represented by z. $\sqrt{x^2 + y^2}$ is called the absolute value of z and is denoted by $|z|$ or $|x + iy|$.

If $x + iy$ is represented by the point (x, y), a "one-to-one correspondence" is established between the complex variable $x + iy$ and the points of the XY-plane. The X-axis is called the axis of reals, the Y-axis the axis of imaginaries. Denoting the polar coordinates of (x, y) by r and θ, $x = r\cos\theta$, $y = r\sin\theta$, and $z = x + iy = r(\cos\theta + r\sin\theta)$, where $r = \sqrt{x^2 + y^2}$, $\theta = \tan^{-1}\frac{y}{x}$. r is the absolute value, and θ is called the amplitude of the complex variable $x + iy$. Hence to the complex variable $x + iy$ there corresponds a straight line determinate in length and direction. A straight line determinate in length and direction is called a vector. Hence there is a "one-to-one correspondence" between the complex variable and plane vectors. As geometric representative of the complex variable may be taken either the point (x, y) or the vector which determines the position of that point with respect to the origin.*

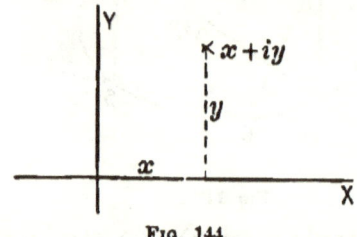

Fig. 144.

* Argand (1806) was the first to represent the complex variable by points in a plane. Gauss (1831) developed the same idea and secured for it a permanent place in mathematics.

Calling a line equal in length to the linear unit and laid off from the origin along the positive direction of the axis of reals the unit vector, the complex variable

$$z = x + iy = r(\cos\theta + i\sin\theta)$$

represents a vector obtained by multiplying the unit vector by the absolute value, then turning the resulting line about its extremity at the origin through an angle equal to the amplitude of the complex variable. When the complex variable is written in the form $r(\cos\theta + i\sin\theta)$, r is the length of the vector,

$$\cos\theta + i\sin\theta$$

the turning factor. In analytic trigonometry it is proved that $\cos\theta + i\sin\theta = e^{i\theta}$.* Hence the complex variable $r(\cos\theta + i\sin\theta) = r \cdot e^{i\theta}$, where the stretching factor (tensor) and turning factor (versor) are neatly separated.

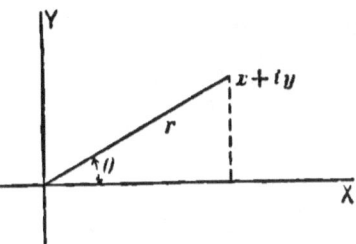

Fig. 145.

Problems. — 1. Locate the points represented by $2 + i5$; $3 - i2$; $-1 + i2$; $i5$; $-i4$; $-3 - i$; $-i7$; $+i7$.

2. Draw the vectors represented by $3 + i2$; $1 - i3$; $-2 + i3$; $-1 - i4$; $-i5$; $3 - i$; $1 + i$.

3. Show that $e^{2n\pi i} = 1$, when n is any integer.

4. Show that $r \cdot e^{i(\theta + 2n\pi)}$ represents the same point for all integral values of n.

5. Locate the different points represented by $e^{\frac{2n\pi i}{3}}$ for integral values of n.

6. Locate the different points represented by $5 \cdot e^{\frac{i(\theta + 2n\pi)}{4}}$ for integral values of n.

* This relation was discovered by Euler (1707-1783).

M

Art. 83. — Arithmetic Operations Applied to Vectors

The sum of two complex variables $x_1 + iy_1$ and $x_2 + iy_2$ is $(x_1 + x_2) + i(y_1 + y_2)$. Hence

$$|(x_1 + iy_1) + (x_2 + iy_2)| = \sqrt{(x_1 + x_2)^2 + (y_1 + y_2)^2},$$

and the amplitude of the sum is $\tan^{-1}\dfrac{y_1 + y_2}{x_1 + x_2}$. The graphic representation shows that the vector corresponding to the sum is found by constructing the vector corresponding to $x_1 + iy_1$ and using the extremity of this vector as origin of a set of new axes parallel to the first axes to construct the vector corresponding to $x_2 + iy_2$. The vector from the origin to the end of the last vector is the vector sum. The vector sum is independent of the order in which the component vectors are constructed. From the figure it is evident that

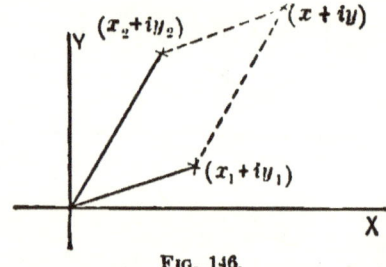

Fig. 146.

$$|(x_1 + iy_1) + (x_2 + iy_2)| \not> |x_1 + iy_1| + |x_2 + iy_2|.$$

The difference between two vectors $x_1 + iy_1$ and $x_2 + iy_2$ is $(x_1 - x_2) + i(y_1 - y_2)$. The graphic representation shows that the vector corresponding to the difference is found by constructing the vector corresponding to $x_1 + iy_1$ and adding to it the vector corresponding to $-x_2 - iy_2$. It is seen that

$$|(x_1 + iy_1) - (x_2 + iy_2)| = \sqrt{(x_1 - x_2)^2 + (y_1 - y_2)^2},$$

the amplitude of the difference is $\tan^{-1}\dfrac{y_1 - y_2}{x_1 - x_2}$, and that the equality of two complex variables requires the equality of the coefficients of the real terms and the imaginary terms separately.

COMPLEX VARIABLE

The product of two complex variables is most readily found by writing these variables in the form $r_1 \cdot e^{i\theta_1}$, $r_2 \cdot e^{i\theta_2}$. The product is $r_1 r_2 \cdot e^{i(\theta_1 + \theta_2)}$, showing that the absolute value of the product is the product of the absolute values of the factors and the amplitude of the product is the sum of the amplitudes of the factors. Hence, writing the complex variables in the form $r_1(\cos\theta_1 + i\sin\theta_1)$, $r_2(\cos\theta_2 + i\sin\theta_2)$, the product is $r_1 r_2[\cos(\theta_1 + \theta_2) + i\sin(\theta_1 + \theta_2)]$, which of course can be shown directly.

Construct the vector corresponding to the multiplier $r_1 \cdot e^{i\theta_1}$ and join its extremity P_1 to the extremity of the unit vector $O1$. Construct the vector corresponding to the multiplicand $r_2 \cdot e^{i\theta_2}$, and on this vector OP_2 as a side homologous to $O1$ construct a triangle OP_2P similar to $OP_1 1$; then OP is the product vector. For, from the similar triangles $OP = r_1 \cdot r_2$, and the angle $XOP = \theta_1 + \theta_2$. The product vector is therefore formed from the vector which is the multiplicand

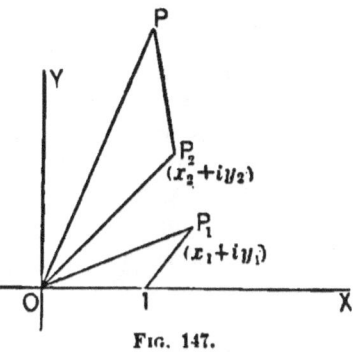

Fig. 147.

in the same manner as the vector which is the multiplier is formed from the unit vector. The product vector is independent of the order of the vector factors and can be zero only when one of the factors is zero.

The quotient of two complex variables $r_1 \cdot e^{i\theta_1}$, $r_2 \cdot e^{i\theta_2}$ is

$$\frac{r_1}{r_2} \cdot e^{i(\theta_1 - \theta_2)};$$

that is, the absolute value of the quotient is the quotient of the absolute values, and the amplitude of the quotient is the amplitude of the dividend minus the amplitude of the divisor.

Construct the vectors OP_1 and OP_2 corresponding to dividend and divisor respectively, and let $O1$ be the unit vector. On

OP_1 as a side homologous to OP_2 construct the triangle OP_1P similar to OP_21, then OP is the quotient vector, for $OP = \dfrac{r_1}{r_2}$ and the angle XOP is $\theta_1 - \theta_2$. The quotient vector is obtained from the vector which is the dividend in the same manner as the unit vector is obtained from the vector which is the divisor.

Extracting the m root of

$$z = r \cdot e^{i\theta} = r \cdot e^{i(\theta + 2n\pi)}$$

there results $z^{\frac{1}{m}} = r^{\frac{1}{m}} \cdot e^{i\left(\frac{\theta}{m} + \frac{n}{m} 2\pi\right)}$.

Since n and m are integers,

$$\frac{n}{m} = q + \frac{r}{m},$$

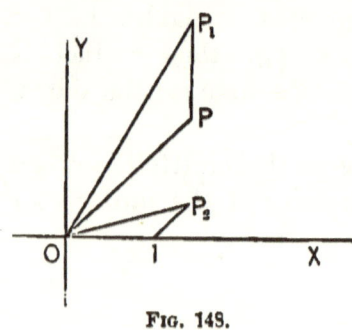

Fig. 148.

where q is an integer and r can have any value from 0 to $m-1$. Hence $z^{\frac{1}{m}} = r^{\frac{1}{m}} \cdot e^{i\left(\frac{\theta}{m} + \frac{r}{m} 2\pi + 2q\pi\right)}$; that is, the m root of z has m values which have the same absolute value and amplitudes differing by $\dfrac{2\pi}{m}$ beginning with $\dfrac{\theta}{m}$.*

Problems. — 1. Add $(2 + i5)$, $(-3 + i2)$, $(5 - i3)$.

2. Find the value of $(3 - i2) + (7 + i4) - (6 - i3)$.
3. Find absolute value and amplitude of
$$(4 - i3) + (2 + i5) - (-3 + i4).$$
4. Construct $(2 - i3) \times (5 + i2) \div (4 - i5)$.
5. Find absolute value and amplitude of $(10 - i7) \times (4 - i3)$.
6. Find absolute value and amplitude of $(15 + i8) \times (5 - i2)$.
7. Construct $(2 + i3)^3$.
8. Construct $(8 - i5)^{\frac{1}{2}}$.
9. Construct $(7 + i4)^{\frac{1}{3}}$.
10. Construct $(9 - i7)^{\frac{1}{4}}$.

* In mechanics coplanar forces, translations, velocities, accelerations, and the moments of couples are vector quantities; that is, quantities which are completely determined by direction and magnitude. Hence the laws of vector combination are the foundation of a complete graphic treatment of mechanics.

COMPLEX VARIABLE

11. Construct the five fifth roots of unity.

12. Construct the roots of $z^2 - 3z + 5 = 0$.

Put $z = x + iy$. There results $(x^2 - y^2 - 3x + 5) + i(2xy - 3y) = 0$. Plot $x^2 - y^2 - 3x + 5 = 0$ and $2xy - 3y = 0$. The values of z determined by the intersections of these curves are the roots of $z^2 - 3z + 5 = 0$.

ART. 84. — ALGEBRAIC FUNCTIONS OF THE COMPLEX VARIABLE

The geometric representative of the real variable is the point system of the X-axis and the geometric representation of a function of a real variable $y = f(x)$ is the line into which this function transforms the X-axis.

The geometric representative of the complex variable is the point system of the XY-plane, and the geometric representation of a function of a complex variable $u + iv = f(x + iy)$ is the system of lines into which this function transforms systems of lines in the XY-plane.

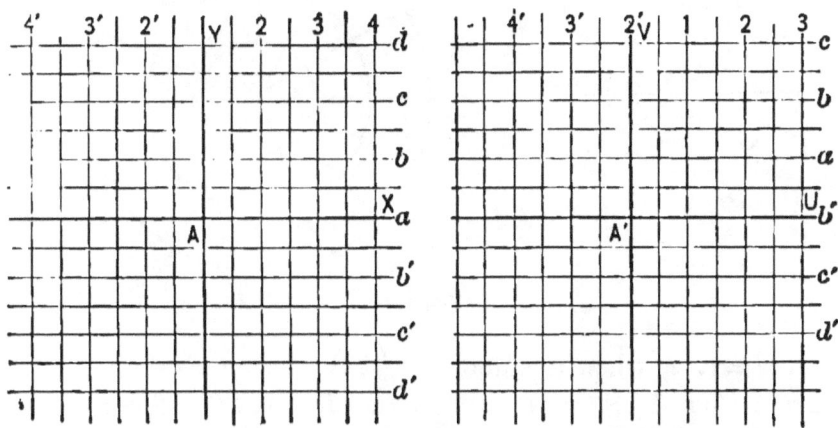

Fig. 149.

When the complex variable is written in the form $x + iy$, it is convenient to use the systems of parallels to the X-axis and to the Y-axis. Take the function $w = z + c$, where w stands for $u + iv$, z for $x + iy$, c for $a + ib$, then

$$u + iv = (x + a) + i(y + b) \text{ and } u = x + a, \ v = y + b.$$

If in the XY-plane a point moves in a parallel to the Y-axis, x is constant, and consequently u is constant. Hence the function $w = z + c$ transforms parallels to the Y-axis into parallels to the V-axis in the UV-plane. In like manner it is shown that $w = z + c$ transforms parallels to the X-axis into parallels to the U-axis. If the variables w and z are interpreted in the same axes, the function $w = z + c$ gives to every point of the XY-plane a motion of translation equal to the translation which carries A to c.

When the complex variable is written in the form $r \cdot e^{i\theta}$, it is convenient to use a system of concentric circles and the system of straight lines through their common center. Take the func-

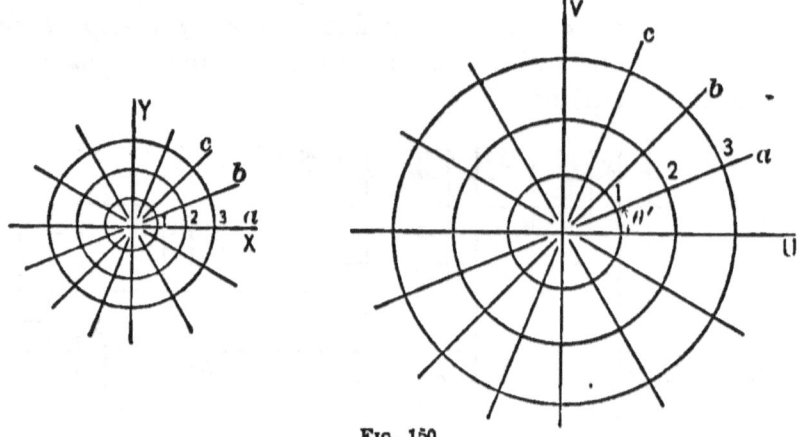

Fig. 150.

tion $w = c \cdot z$, when w stands for $R \cdot e^{i\Theta}$, z for $r \cdot e^{i\theta}$, and c for $r' \cdot e^{i\theta'}$; then $R \cdot e^{i\Theta} = rr' \cdot e^{i(\theta+\theta')}$, and $R = rr'$, $\Theta = \theta + \theta'$. If a point in the XY-plane describes the circumference of a circle center at origin, r is constant, and consequently R is constant, and the corresponding point describes a circumference in the UV-plane, center at origin, and radius r' times the radius of the corresponding circle in the XY-plane. If the point in the XY-plane moves in a straight line through the origin, θ is constant, and consequently Θ is constant, and the corresponding point in the UV-plane moves in a straight line through the origin. If the variables w and z are interpreted in the same axes X and Y, the function $w = c \cdot z$ either stretches the XY-plane outward from the origin, or shrinks it toward the origin, according as r' is greater or less than unity, and then turns the whole plane about the origin through the angle θ'.

The function $w = \dfrac{1}{z}$ may be written $R \cdot e^{i\Theta} = \dfrac{1}{r} e^{-i\theta}$, whence $R = \dfrac{1}{r}$, $\Theta = -\theta$. A circle in the XY-plane with center at the origin is transformed into a circle in the UV-plane with center at the origin, the radius of one circle being the reciprocal of the radius of the other. A straight line through the origin in

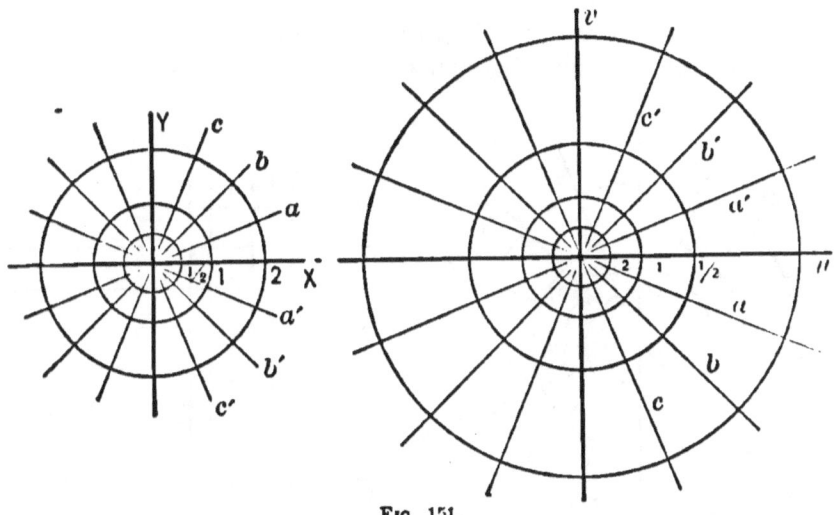

Fig. 151.

the XY-plane making an angle θ with the X-axis, is transformed into a straight line through the origin in the UV-plane making an angle $-\theta$ with the U-axis. If w and z are interpreted in the same axes, the function $w = \dfrac{1}{z}$ is equivalent to a transformation by reciprocal radii vectors with respect to the unit circle, and a transformation by symmetry with respect to the axis of reals.

In the equation $w = z^3$, or $R \cdot e^{i\Theta} = r^3 \cdot e^{i3\theta}$, w is a single valued function of z, but z is a three-valued function of w. Since $r = R^{\frac{1}{3}}$, $\theta = \frac{\Theta}{3} + \frac{2n\pi}{3}$, the absolute values of the three values of z are the same, but their amplitudes differ by 120°. The positive half of the U-axis, $\theta = 0$, corresponds to the positive half of the X-axis, and the lines through the origin making angles of 120° and 240° with the X-axis. The entire UV-plane is pictured by the function $w = z^3$ on each of the three parts into which these lines divide the XY-plane.

Art. 85. — Generalized Transcendental Functions

Since $z = x + iy = r \cdot e^{i(\theta + 2n\pi)}$, $\log z = \log r + i(\theta + 2n\pi)$. The equation $w = \log z$ may be written $u + iv = \log r + i(\theta + 2n\pi)$.

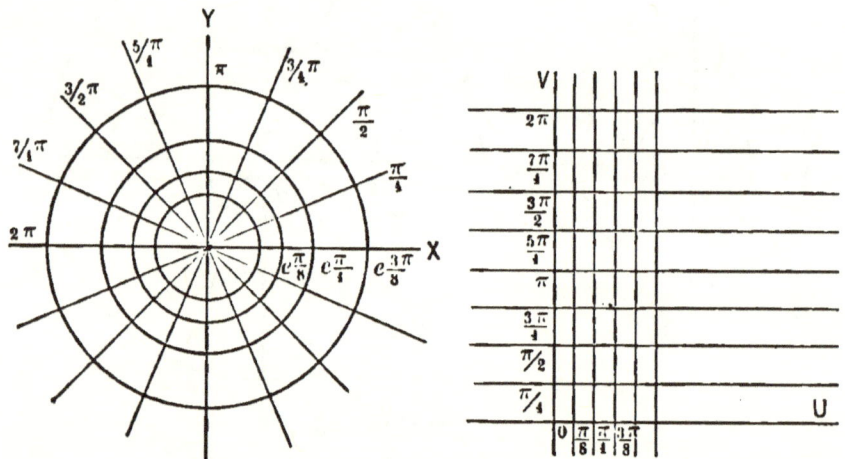

Fig. 152.

Hence $u = \log r$, $v = \theta + 2n\pi$. To the circle $r =$ constant in the XY-plane there corresponds in the UV-plane a straight line parallel to the V-axis; to the straight line $\theta =$ constant in the XY-plane there corresponds in the UV-plane a system of parallels to the U-axis at distances of 2π from one another; w is an infinite valued function of z, but z is a single valued function of w. The entire XY-plane is pictured between any two successive parallels to the U-axis at distances of 2π.

Writing the function $w = \sin(x + iy)$ in the form
$$u + iv = \sin x \cos iy + \cos x \sin iy,$$
and remembering that
$$\cosh y = \tfrac{1}{2}(e^y + e^{-y}) = \cos iy, \quad \sinh y = \tfrac{1}{2}(e^y - e^{-y}) = -i \sin iy,$$
there results
$$u + iv = \cosh y \sin x - i \sinh y \cos x,$$
whence
$$u = \cosh y \sin x, \quad v = -\sinh y \cos x, \quad \text{and} \quad \sin x = \frac{u}{\cosh y},$$
$$\cos x = \frac{-v}{\sinh y}, \quad \cosh y = \frac{u}{\sin x}, \quad \sinh y = -\frac{v}{\cos x}.$$

Substituting in $\sin^2 x + \cos^2 x = 1$ and $\cosh^2 y - \sinh^2 y = 1$, there results
$$\frac{u^2}{\cosh^2 y} + \frac{v^2}{\sinh^2 y} = 1, \quad \frac{u^2}{\sin^2 x} - \frac{v^2}{\cos^2 x} = 1.$$

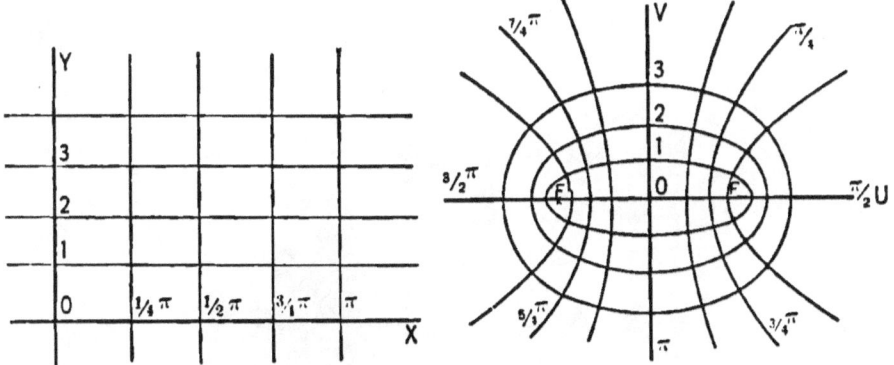

Fig. 153.

These equations when x and y are respectively constant represent a system of confocal conic sections with the foci at $(+1, 0), (-1, 0)$. The entire system of ellipses filling up the UV-plane is obtained by assigning to y values from $+\infty$ to $-\infty$; the entire system of hyperbolas filling up the UV-plane is obtained by assigning to x values from 0 to 2π. Hence $w = \sin(x + iy)$ pictures that part of the XY-plane between two parallels to the Y-axis at a distance of 2π from each other on the entire UV-plane.*

Problems. — 1. Show that $w = \dfrac{1}{z}$ transforms the system of straight lines through $a + ib$, and the system of circles concentric at this point into systems of orthogonal circles.

2. Find what part of the XY-plane is transformed into the entire UV-plane by the function $w = z^2$.

3. Into what systems of lines does $w = \cos z$ transform the parallels to the X-axis and to the Y-axis?

* The geometric treatment of functions of the complex variable has been extensively developed by Riemann (1826–66) and his school.

ANALYTIC GEOMETRY OF THREE DIMENSIONS

CHAPTER XIII

POINT, LINE, AND PLANE IN SPACE

Art. 86. — Rectilinear Space Coordinates

Through a point in space draw any three straight lines not in the same plane. The point is called the origin of coordinates, the lines the axes of coordinates, the planes determined by the lines taken two and two, the coordinate planes. The distance of any point P from a coordinate plane is measured on a parallel to that axis which does not lie in the plane, and the direction of the point from the plane is denoted by the algebraic sign prefixed to the number expressing the distance. The interpretation of these signs is indicated in the figure. If the distance and direction of the point from the YZ-plane is given, $x = a$, the

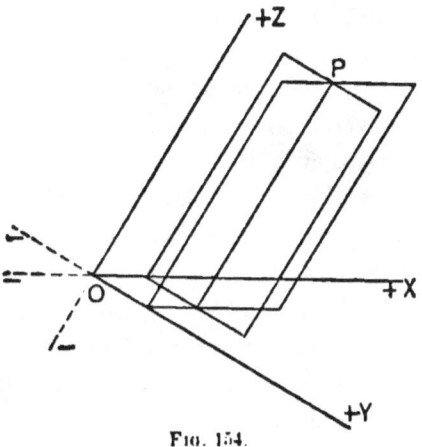

Fig. 154.

point must lie in a determinate plane parallel to the YZ-plane. If the distance and direction of the point from the XZ-plane is given, $y = b$, the point must lie in a determinate plane parallel to the XZ-plane. If it is known that $x = a$ and $y = b$, the point must lie in each of two planes parallel, the one to the

YZ-plane, the other to the XZ-plane, and therefore the point must lie in a determinate straight line parallel to the Z-axis. If the distance and direction of the point from the XY-plane $z = c$ is also given, the point must lie in a determinate plane parallel to the XY-plane and in a determinate line parallel to the Z-axis; that is, the point is completely determined.

Conversely, to every point in space there corresponds one, and only one, set of values of the distances and directions of the point from the coordinate planes. For through the given point only one plane can be passed parallel to a coordinate plane, a fact which determines a single value for the distance and direction of the point from that coordinate plane.

The point whose distances and directions from the coordinate planes are represented by x, y, z is denoted by the symbol (x, y, z), and x, y, z are called the rectilinear coordinates of the point. There is seen to be a "one-to-one correspondence" between the symbol (x, y, z) and the points of space.

Observe that $x = a$ interpreted in the ZX-plane represents a straight line parallel to the Z-axis; interpreted in the XY-plane a straight line parallel to the Y-axis; but when interpreted in space it represents the plane parallel to the YZ-plane containing these two lines. The equations $x = a$, $y = b$ interpreted in the XY-plane represents a point; interpreted in space they represent a straight line through this point parallel to the Z-axis.

If the axes are perpendicular to each other, the coordinates are called rectangular, in all other cases oblique.

Problems. — **1.** Write the equation of the plane parallel to the YZ-plane cutting the X-axis 5 to the right of the origin.

2. What is the equation of the YZ-plane?

3. What is the locus of the points at a distance 7 below the XY-plane? Write equation of locus.

4. Write the equations of the line parallel to the X-axis at a distance $+ 5$ from the XY-plane and at a distance $- 5$ from the XZ-plane.

POINT, LINE, AND PLANE IN SPACE

5. Write the equations of the origin.

6. What are the coordinates of the point on the Z-axis 10 below the XY-plane?

7. What are the equations of the Z-axis?

8. What are the equations of a line parallel to the Z-axis?

9. Explain the limitations of the position of a point imposed by placing $x = +5$, then $y = -5$, then $z = -3$.

10. Locate the points $(2, -3, 5)$; $(-2, 3, -5)$.

11. Locate $(0, 4, 5)$; $(2, 0, -3)$.

12. Locate $(0, 0, -5)$; $(0, -5, 0)$.

13. Show that (a, b, c), $(-a, b, c)$ are symmetrical with respect to the YZ-plane.

14. Show that (a, b, c), $(-a, -b, c)$ are symmetrical with respect to the Z-axis.

15. Show that (a, b, c), $(-a, -b, -c)$ are symmetrical with respect to the origin.

ART. 87. — POLAR SPACE COORDINATES

Let (x, y, z) be the rectangular coordinates of any point P in space. Call the distance from the origin O to the point r, the angle made by OP with its projection OP' on the XY-plane θ, the angle made by the projection OP' with the X-axis ϕ. r, ϕ, θ are the polar coordinates of the point P. From the figure

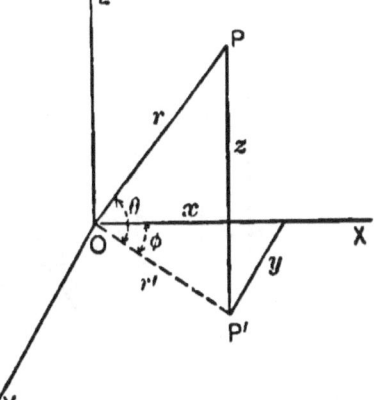

Fig. 155.

$OP' = r \cdot \cos \theta$,

$x = OP' \cdot \cos \phi = r \cos \theta \cos \phi$,

$y = OP' \cdot \sin \phi = r \cos \theta \sin \phi$,

$z = r \sin \theta$,

formulas which express the rectangular coordinates of any point in space in terms of the polar coordinates of the same point.

From the figure are also obtained $r = (x^2 + y^2 + z^2)^{\frac{1}{2}}$, $\sin \theta = \dfrac{z}{r}$, $\tan \phi = \dfrac{y}{x}$, formulas which express the polar coordinates of any point in space in terms of the rectangular coordinates of the same point.

Problems. — 1. Locate the points whose polar coordinates are 5, 15°, 60°; 8, 90°, 45°.

2. Find the polar coordinates of the point (3, 4, 5).

3. Find the rectangular coordinates of the point (10, 30°, 60°).

4. Find the distance from the origin to the point (4, 5, 7).

Art. 88. — Distance between Two Points

Let the rectangular coordinates of the points be (x', y', z'), (x'', y'', z''). From the figure

$$D^2 = D'^2 + (z' - z'')^2, \quad D'^2 = (x' - x'')^2 + (y' - y'')^2,$$

hence (1) $\quad D^2 = (x' - x'')^2 + (y' - y'')^2 + (z' - z'')^2.$

$x' - x''$ is the projection of D on the X-axis; $y' - y''$ the pro-

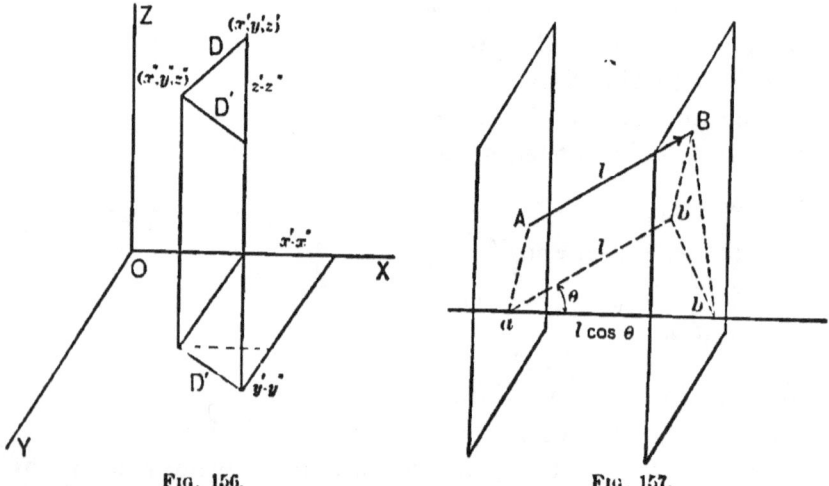

Fig. 156.   Fig. 157.

jection of D on the Y-axis; $z' - z''$ the projection of D on the Z-axis.* Calling the angles which D makes with the coordinate axes respectively X, Y, Z,

$$x' - x'' = D \cos X, \; y' - y'' = D \cos Y, \; z' - z'' = D \cos Z.$$

Substituting in (1), there results

$$D^2 \cos^2 X + D^2 \cos^2 Y + D^2 \cos^2 Z = D^2,$$

whence
$$\cos^2 X + \cos^2 Y + \cos^2 Z = 1;$$

that is, the sum of the squares of the cosines of the three angles which a straight line in space makes with the rectangular coordinate axes is unity.

The distance from (x', y', z') to the origin is $\sqrt{x'^2 + y'^2 + z'^2}$. If the point (x, y, z) moves so that its distance from (x', y', z') is always R, the locus of the point is the surface of a sphere and $(x - x')^2 + (y - y'^2) + (z - z')^2 = R^2$, which expresses the geometric law governing the motion of the point, is the equation of the sphere whose center is (x', y', z'), radius R.

Problems. — 1. Find distance of $(2, -3, 5)$ from origin.

2. Find the angles which the line from $(3, 4, 5)$ to the origin makes with the coordinate axes.

3. Find distance between points $(-2, 4, -5)$, $(3, -4, 5)$.

4. Write equation of locus of points whose distance from $(4, -1, 3)$ is 5.

5. Write equation of sphere center at origin, $(2, 1, -3)$ on surface.

6. The locus of points equidistant from (x', y', z'), (x'', y'', z'') is the plane bisecting at right angles the line joining these points. Find the equation of the plane.

7. Find the equation of the plane bisecting at right angles the line joining $(2, 1, 3)$, $(4, 3, -2)$.

---

* The projection of one straight line in space on another is the part of the second line included between planes through the extremities of the first line perpendicular to the second. The projection is given in direction and magnitude by the product of the line to be projected into the cosine of the included angle.

8. Show that $\left(\dfrac{x'+x''}{2},\ \dfrac{y'+y''}{2},\ \dfrac{z'+z''}{2}\right)$ is the point midway between (x', y', z'), (x'', y'', z'').

9. Find the point midway between $(4, 5, 7)$, $(2, -1, 3)$.

10. Find the equation of the sphere which has the points $(4, 5, 8)$, $(2, -3, 4)$ at the extremities of a diameter.

11. Write the equation of the sphere with the origin on the surface, center $(5, 0, 0)$.

12. Find angles which the line through $(2, 3, -5)$, $(4, -2, 3)$ makes with the coordinate axes.

13. The length of the line from the origin to (x, y, z) is r, the line makes with the axes the angles α, β, γ. Show that $x = r\cos\alpha$, $y = r\cos\beta$, $z = r\cos\gamma$.

Art. 89. — Equations of Lines in Space

Suppose any line in space to be given. From every point of the line draw a straight line perpendicular to the XZ-plane. There is formed the surface which projects the line in space on the XZ-plane. The values of x and z are the same for all points in the straight line which projects a point of the line in space on the XZ-plane. Hence the equation of the projection of the line in space on the XZ-plane when interpreted in space represents the projecting surface. The projection of the line in space on the XZ-plane determines one surface on which the line in space must lie.

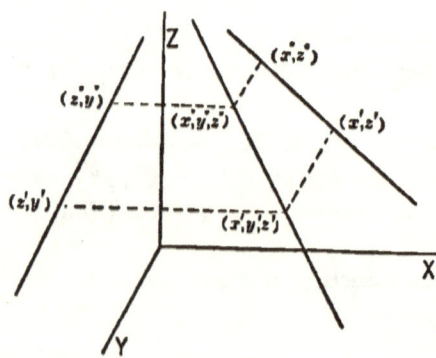

Fig. 158.

The projection of the line in space on the YZ-plane determines a second surface on which the line in space must lie. The equations of the projections of the line in space on the coordinate planes XZ and

POINT, LINE, AND PLANE IN SPACE 177

YZ therefore determine the line in space and are called the equations of the line in space. By eliminating z from the equations of the projections of the line on the planes XZ and YZ, the equation of the projection of the line on the XY-plane is found.

Art. 90. — Equations of the Straight Line

The equations of the projections of the straight line on the coordinate planes XZ and YZ are $x = az + \alpha$, $y = bz + \beta$. The geometric meaning of a, b, α, β is indicated in the figure. The elimination of z gives

$$y - \beta = \frac{b}{a}(x - \alpha),$$

the equation of the projection of the line in the XY-plane.

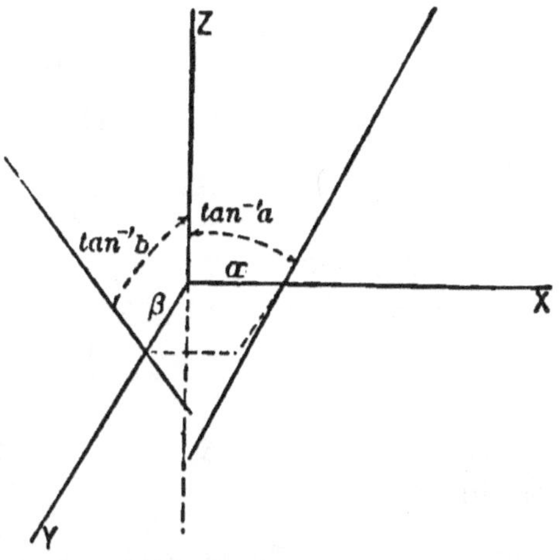

Two points,

(x', y', z'), (x'', y'', z''),

determine a straight line in space. The projection of the line through the

Fig. 159.

points (x', y', z'), (x'', y'', z'') on the ZX-plane is determined by the projections (x', z'), (x'', z'') of the points on the ZX-plane, likewise the projection of the line on the ZY-plane is determined by the points (z', y'), (z'', y''). Hence the equations of the straight lines through (x', y', z'), (x'', y'', z'') are

$$x - x' = \frac{x' - x''}{z' - z''}(z - z'), \quad y - y' = \frac{y' - y''}{z' - z''}(z - z').$$

A straight line is also determined by one point and the direction of the line. Let (x', y', z') be one point of the line, α, β, γ

the angles which the line makes with the axes X, Y, Z respectively. Let (x, y, z) be any point of the line, d its distance from (x', y', z'). Then

$$\frac{x-x'}{\cos\alpha} = \frac{y-y'}{\cos\beta} = \frac{z-z'}{\cos\gamma} = d \quad (1)$$

is the equation of the line. This equation is equivalent to the equations $x = x' + d\cos\alpha$, $y = y' + d\cos\beta$, $z = z' + d\cos\gamma$, which express the coordinates of any point of the line in terms of the single variable d.

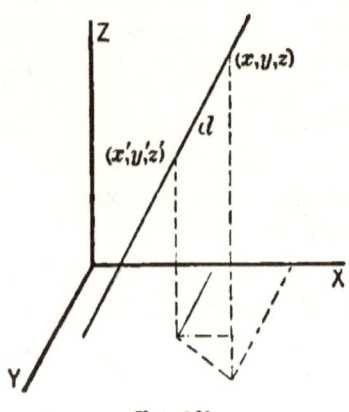

Fig. 160.

If the straight line (1) contains the point (x'', y'', z''),

$$\frac{x''-x'}{\cos\alpha} = \frac{y''-y'}{\cos\beta} = \frac{z''-z'}{\cos\gamma}. \quad (2)$$

Eliminate $\cos\alpha$, $\cos\beta$, $\cos\gamma$ from (1) and (2) by division, and the equation of the straight line through two points is obtained

$$\frac{x-x'}{x''-x'} = \frac{y-y'}{y''-y'} = \frac{z-z'}{z''-z'},$$

as found before. α, β, γ are called the direction angles of the straight line.

Problems. — 1. The projections of a straight line on the planes XZ and YZ are $2x + 3z = 5$, $\frac{y}{2} - \frac{z}{3} = 1$. Find the projection on the XY plane.

2. Find the intersections of $x = 3z + 5$, $y = 2z - 3$ with the coordinate planes.

3. Write the equations of the straight line through $(2, 3, 1)$, $(-1, 3, 5)$.

4. Write the equations of the straight line through the origin and the point $(4, -1, 2)$.

5. Write the equations of the straight line through $(3, 1, 2)$ whose direction angles are $(60°, 45°, 60°)$.

6. The direction angles of a straight line are $(45°, 60°, 60°)$; $(4, 5, 6)$ is a point of the line. Find the coordinates of the point 10 from $(4, 5, 6)$.

Art. 91. — Angle between Two Straight Lines

Let $\dfrac{x-a}{\cos\alpha}=\dfrac{y-b}{\cos\beta}=\dfrac{z-c}{\cos\gamma}$, $\dfrac{x-a'}{\cos\alpha'}=\dfrac{y-b'}{\cos\beta'}=\dfrac{z-c'}{\cos\gamma'}$

be the straight lines. The angle between the lines is by definition the angle between parallels to the lines through the origin. Let OM' and OM'' be these parallels through the origin. From any point $P'(x', y', z')$ of OM' draw a perpendicular $P'P''$ to OM''. Then OP'' is the projection of OP' on OM'', and OP'' is also the projection of the broken line $(x'+y'+z')$ on OM''.* Hence

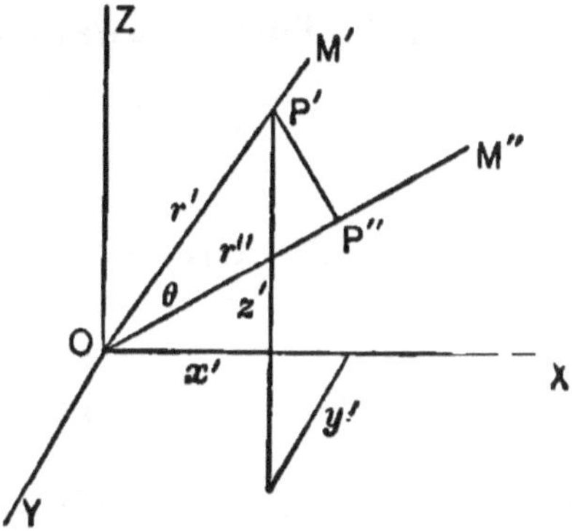

$r' \cos\theta = x' \cos\alpha' + y' \cos\beta' + z' \cos\gamma$,

$$\cos\theta = \frac{x'}{r'}\cos\alpha' + \frac{y'}{r'}\cos\beta' + \frac{z'}{r'}\cos\gamma', \qquad (1)$$

that is $\cos\theta = \cos\alpha \cos\alpha' + \cos\beta \cos\beta' + \cos\gamma \cos\gamma'$.

Fig. 161.

* The sum of the projections of the parts of a broken line on any straight line is the part of the line included between the projections of the extremities of the broken line. ab is the projection of AB; bc is the projection of BC; ac is the projection of $AB + BC$.

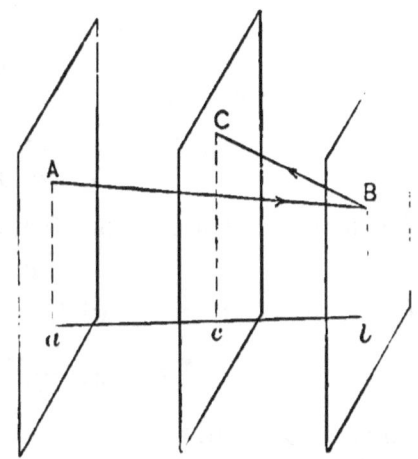

Fig. 162.

If the equations of the lines are written in the form
$$x = az + \alpha,\ y = bz + \beta;\ x = a'z + \alpha',\ y = b'z + \beta',$$
the equations of parallel lines through the origin are
$$x = az,\ y = bz;\ x = a'z,\ y = b'z.$$
Let (x', y', z') be any point of the first line, its distance from the origin r'. Then $x' = az'$, $y' = bz'$, $r'^2 = x'^2 + y'^2 + z'^2$, whence

$$\cos \alpha = \frac{x'}{r'} = \frac{a}{\sqrt{1 + a^2 + b^2}},$$

$$\cos \beta = \frac{y'}{r'} = \frac{b}{\sqrt{1 + a^2 + b^2}},$$

$$\cos \gamma = \frac{z'}{r'} = \frac{1}{\sqrt{1 + a^2 + b^2}}.$$

Likewise if (x'', y'', z'') is any point of the second line, r'' its distance from the origin,

$$\cos \alpha' = \frac{x''}{r''} = \frac{a'}{\sqrt{1 + a'^2 + b'^2}},$$

$$\cos \beta' = \frac{y''}{r''} = \frac{b'}{\sqrt{1 + a'^2 + b'^2}},$$

$$\cos \gamma' = \frac{z''}{r''} = \frac{1}{\sqrt{1 + a'^2 + b'^2}}.$$

Substituting in (1),
$$\cos \theta = \frac{1 + aa' + bb'}{\sqrt{1 + a^2 + b^2}\sqrt{1 + a'^2 + b'^2}}.$$

When the lines are perpendicular, $\cos \theta = 0$, whence
$$1 + aa' + bb' = 0.$$
When the lines are parallel, $\cos \theta = 1$, whence
$$1 = \frac{1 + aa' + bb'}{\sqrt{1 + a^2 + b^2}\sqrt{1 + a'^2 + b'^2}},$$
which reduces to
$$(a' - a)^2 + (b' - b)^2 + (ab' - a'b)^2 = 0.$$

This equation requires that $a = a'$, $b = b'$; that is, if two lines are parallel, their projections on the coordinate planes are parallel.

The equations of the straight line through (x', y', z') parallel to $x = az + \alpha$, $y = bz + \beta$ are $x - x' = a(z - z')$, $y - y' = b(z - z')$.

The straight line (1) $x - x' = a'(z - z')$, $y - y' = b'(z - z')$ through the point (x', y', z') is perpendicular to the straight line (2) $x = az + \alpha$, $y = bz + \beta$ when a' and b' satisfy the equation $1 + aa' + bb' = 0$. This equation is satisfied by an infinite number of pairs of values of a' and b'. This is as it ought to be, for through the given point a plane can be passed perpendicular to the given line, and every line in this plane is perpendicular to the given line, and conversely. Hence if the straight line (1) is governed in its motion by the equation $1 + aa' + bb' = 0$, it generates the plane through (x', y', z') perpendicular to the straight line (2). $1 + aa' + bb' = 0$ is the line equation of the plane.

To find the relation between the constants in the equations of two straight lines $x = az + \alpha$, $y = bz + \beta$, $x = a'z + \alpha'$, $y = b'z + \beta'$, which causes the lines to intersect, make these equations simultaneous and solve the equations of the projections on the XZ-plane, also the equations of the projections on the YZ-plane, for z. The two values of z, $\dfrac{\alpha' - \alpha}{a - a'}$ and $\dfrac{\beta' - \beta}{b - b'}$ must be equal if the lines intersect. Hence for intersection the equation $(a - a')(\beta' - \beta) - (b - b')(\alpha' - \alpha) = 0$ must be satisfied, and the coordinates of the point of intersection are $x = \dfrac{a\alpha' - a'\alpha}{a - a'}$, $y = \dfrac{b\beta' - b'\beta}{b - b'}$, $z = \dfrac{\alpha' - \alpha}{a - a'}$. When $a = a'$ and $b = b'$, the point of intersection is at infinity, and the lines are parallel, as found before.

Problems. — 1. Find the angle between the lines
$x = 3z + 1$, $y = -2z + 5$; $x = z + 2$, $y = -z + 4$.

2. Find the angle between the lines through $(1, 1, 2)$, $(-3, -2, 4)$ and $(2, 1, -2)$, $(3, 2, 1)$.

3. Find equations of line through $(4, -2, 3)$ parallel to $x = 4z + 1$, $y = 2z - 5$.

4. Find line through $(1, -2, 3)$ intersecting $x = -2z + 3$, $y = z + 5$ at right angles.

5. Find distance from $(2, 2, 2)$ to line $x = 2z + 1$, $y = -2z + 3$.

6. Find equations of line intersecting each of the lines $x = 3z + 4$, $y = -z + 2$ and $y = 2z - 5$, $x = -z + 2$ at right angles.

7. For what value of a do the lines $x = 3z + a$, $y = 2z + 5$ and $x = -2z - 3$, $y = 4z - 5$ intersect?

8. Find the equations of the straight line through the origin intersecting at right angles the line through $(4, 2, -1)$, $(1, 2, -3)$.

9. Find distance of point of intersection of lines $x = 2z + 1$, $y = 2z + 2$ and $x = z + 5$, $y = 4z - 6$ from origin.

10. Find distance from origin to line $x = 4z - 5$, $y = -2z + 3$.

Art. 92. — The Plane

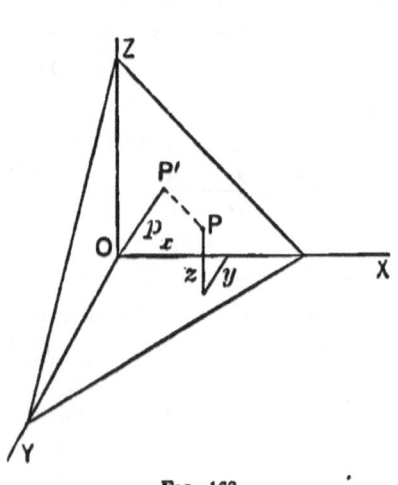

Fig. 163.

A plane is determined when the length and direction of the perpendicular from the origin to the plane are given. Call the length of the perpendicular p, the direction angles of the perpendicular α, β, γ. Let $P(x, y, z)$ be any point in the plane. The projection of the broken line $(x + y + z)$ on the perpendicular OP equals p for all points in the plane and for no others. Hence $x \cos \alpha + y \cos \beta + z \cos \gamma = p$ is the equation of the plane. This is called the normal equation of the plane.

Every first degree equation in three variables when interpreted in rectangular coordinates represents a plane. The locus represented by $Ax + By + Cz + D = 0$ is the same as the locus represented by $x\cos\alpha + y\cos\beta + z\cos\gamma - p = 0$ if

$$\frac{\cos\alpha}{A} = \frac{\cos\beta}{B} = \frac{\cos\gamma}{C} = -\frac{p}{D}.$$

Combining with

$$\cos^2\alpha + \cos^2\beta + \cos^2\gamma = 1, \ \cos\alpha = \frac{A}{\sqrt{A^2 + B^2 + C^2}},$$

$$\cos\beta = \frac{B}{\sqrt{A^2 + B^2 + C^2}}, \ \cos\gamma = \frac{C}{\sqrt{A^2 + B^2 + C^2}},$$

$$p = -\frac{D}{\sqrt{A^2 + B^2 + C^2}}.$$

Hence the factor $\dfrac{1}{\sqrt{A^2 + B^2 + C^2}}$ transforms $Ax + By + Cz + D = 0$ into an equation of the form $x\cos\alpha + y\cos\beta + z\cos\gamma = p$, which is the equation of a plane.

$\dfrac{x}{a} + \dfrac{y}{b} + \dfrac{z}{c} = 1$ is the equation of the plane whose intercepts on the coordinate axes are a, b, c. This is the intercept equation of the plane.

The plane represented by the equation $Ax + By + Cz + D = 0$ depends on the relative values of the coefficients. Hence the equation of the plane has three parameters. To find the equation of the plane through three points (x', y', z'), (x'', y'', z''), (x''', y''', z'''), substitute these coordinates for x, y, z in (1) $A'x + B'y + C'z + 1 = 0$, solve the resulting equations for A', B', C', and substitute in (1).

The intersections of a plane with the coordinate planes are called the traces of the plane on the coordinate planes. The equation of the trace of $Ax + By + Cz + D = 0$ on the XZ-plane is found by making $y = 0$ in the equation of the plane. The trace is therefore $Ax + Cz + D = 0$. The trace on YZ is $By + Cz + D = 0$, on XY is $Ax + By + D = 0$.

184 ANALYTIC GEOMETRY

For points in the intersection of the planes

$$Ax + By + Cz + D = 0 \text{ and } A'x + B'y + C'z + D' = 0$$

these equations are simultaneous. Eliminating y,

$$(AB' - A'B)x + (CB' - C'B)z + (DB' - D'B) = 0,$$

the equation of the projection of the intersection on the coordinate plane XZ. In like manner the equations of the projections of the intersection on the planes YZ and XY are obtained.

Problems. — 1. Write the equation of the plane whose intercepts on the axes are 2, −4, −3.

2. Find the equation of the plane through (2, −3, 4) perpendicular to the line joining this point to the origin.

3. Find the equation of the plane through (2, 5, 1), (3, 2, −5), (1, −3, 7).

4. Find the equations of the traces of $3x - y + 5z - 15 = 0$.

5. Find the equations of the intersection of $3x + 5y - 7z + 10 = 0$, $5x - 14y + 3z - 15 = 0$.

6. Find the equation of the plane through (3, −2, 5) perpendicular to $\frac{x-1}{\cos 60°} = \frac{y+2}{\cos 45°} = \frac{z-3}{\cos 60°}$.

7. Find the direction angles of a perpendicular to the plane

$$2x - 3y + 5z = 6.$$

8. Find the length of the perpendicular from the origin to

$$2x - 3y + 5z = 6.$$

Art. 93. — Distance from a Point to a Plane

Let (x', y', z') be a given point, $x \cos \alpha + y \cos \beta + z \cos \gamma = p$, a given plane. Through (x', y', z') pass a plane parallel to the given plane. The equation of this parallel plane is

$$x \cos \alpha + y \cos \beta + z \cos \gamma = OP'''.$$

The point (x', y', z') lies in this plane, therefore
$$x' \cos \alpha + y' \cos \beta + z' \cos \gamma = OP''.$$
Subtracting OP from both sides of this equation,
$$x' \cos \alpha + y' \cos \beta + z' \cos \gamma - p = PP'';$$
that is, the perpendicular distance from (x', y', z') to
$$x \cos \alpha + y \cos \beta + z \cos \gamma - p = 0$$
is the left-hand member of this equation evaluated for (x', y', z'). The sign of the perpendicular is plus when the origin and the point (x', y', z') are on different sides of the plane, minus when the origin and the point (x', y', z') are on the same side of the plane.

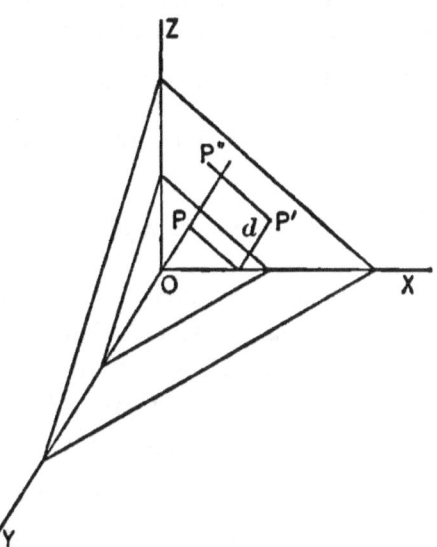

Fig. 164.

The distance from (x', y', z') to the plane $Ax + By + Cz + D = 0$ is found by transforming the equation of the plane into the form $x \cos \alpha + y \cos \beta + z \cos \gamma - p = 0$ to be
$$\frac{Ax' + By' + Cz' + D}{\sqrt{A^2 + B^2 + C^2}}.$$

Let $x \cos \alpha + y \cos \beta + z \cos \gamma - p = 0$ and
$$x \cos \alpha' + y \cos \beta' + z \cos \gamma' - p' = 0$$
be the faces of a diedral angle.
$$(x \cos \alpha + y \cos \beta + z \cos \gamma - p) \pm (x \cos \alpha' + y \cos \beta' + z \cos \gamma' - p') = 0$$
is the equation of the locus of points equidistant from the faces; that is, the equation of the bisectors of the diedral angle.

Problems. — 1. Find distance from origin to plane
$$11x - 13y + 17z + 22 = 0.$$

2. Find distance from $(3, -2, 7)$ to $3x + 7y - 10z + 5 = 0$.

3. Write the equations of the bisectors of the diedral angles whose faces are $2x + 5y - 7z = 10$, and $4x - y + 6z - 15 = 0$.

4. Find distance from $(0, 5, 7)$ to $\frac{x}{2} + \frac{y}{3} + \frac{z}{4} = 1$.

5. Find distance from origin to $\frac{2}{3}x - \frac{4}{5}y + \frac{5}{6}z = 1$.

ART 94. — ANGLE BETWEEN TWO PLANES

Let $x \cos\alpha + y \cos\beta + z \cos\gamma = p$, $x \cos\alpha' + y \cos\beta' + z \cos\gamma' = p'$ be two given planes, θ their included angle. The angle between the planes is the angle between the perpendiculars to the planes from the origin. Hence

$$\cos\theta = \cos\alpha \cos\alpha' + \cos\beta \cos\beta' + \cos\gamma \cos\gamma'.$$

If the equations of the planes are in the form
$$Ax + By + Cz + D = 0, \quad A'x + B'y + C'z + D' = 0,$$

$$\cos\alpha = \frac{A}{\sqrt{A^2 + B^2 + C^2}}, \qquad \cos\alpha' = \frac{A'}{\sqrt{A'^2 + B'^2 + C'^2}},$$

$$\cos\beta = \frac{B}{\sqrt{A^2 + B^2 + C^2}}, \qquad \cos\beta' = \frac{B'}{\sqrt{A'^2 + B'^2 + C'^2}},$$

$$\cos\gamma = \frac{C}{\sqrt{A^2 + B^2 + C^2}}, \qquad \cos\gamma' = \frac{C'}{\sqrt{A'^2 + B'^2 + C'^2}}.$$

Hence $\cos\theta = \dfrac{AA' + BB' + CC'}{\sqrt{A^2 + B^2 + C^2}\sqrt{A'^2 + B'^2 + C'^2}}.$

The planes are perpendicular when $AA' + BB' + CC' = 0$; parallel when $1 = \dfrac{AA' + BB' + CC'}{\sqrt{A^2 + B^2 + C^2}\sqrt{A'^2 + B'^2 + C'^2}}$, which reduces to $(AB' - A'B)^2 + (AC' - A'C)^2 + (BC' - B'C)^2 = 0$, whence $\dfrac{A}{A'} = \dfrac{B}{B'} = \dfrac{C}{C'}.$

The angle between the plane $x\cos\alpha + y\cos\beta + z\cos\gamma = p$ and the line $\dfrac{x-x'}{\cos\alpha'} = \dfrac{y-y'}{\cos\beta'} = \dfrac{z-z'}{\cos\gamma'}$ is the complement of the angle between the line and the perpendicular to the plane. Hence $\sin\theta = \cos\alpha\cos\alpha' + \cos\beta\cos\beta' + \cos\gamma\cos\gamma'$. If the equations of line and plane are in the form $x = az + \alpha$, $y = bz + \beta$, and $Ax + By + Cz + D = 0$,

$$\cos\alpha = \frac{A}{\sqrt{A^2+B^2+C^2}}, \quad \cos\alpha' = \frac{a}{\sqrt{1+a^2+b^2}},$$

$$\cos\beta = \frac{B}{\sqrt{A^2+B^2+C^2}}, \quad \cos\beta' = \frac{b}{\sqrt{1+a^2+b^2}},$$

$$\cos\gamma = \frac{C}{\sqrt{A^2+B^2+C^2}}, \quad \cos\gamma' = \frac{1}{\sqrt{1+a^2+b^2}}.$$

Hence $\quad \sin\theta = \dfrac{Aa + Bb + C}{\sqrt{A^2+B^2+C^2}\sqrt{1+a^2+b^2}}.$

The line is parallel to the plane when $Aa + Bb + C = 0$; perpendicular when $1 = \dfrac{Aa + Bb + C}{\sqrt{A^2+B^2+C^2}\sqrt{1+a^2+b^2}}$, which reduces to $(Ab - Ba)^2 + (A - Ca)^2 + (B - Cb)^2 = 0$, whence $a = \dfrac{A}{C}$, $b = \dfrac{B}{C}$.

To find the intersection of the line $x = az + \alpha$, $y = bz + \beta$, and the plane $Ax + By + Cz + D = 0$, make these equations simultaneous, and solve for x, y, z. There results

$$z = -\frac{A\alpha + B\beta + D}{Aa + Bb + C}.$$

If $Aa + Bb + C = 0$, the point of intersection goes to infinity, and the line and plane are parallel, as found before. If $A\alpha + B\beta + D$ also vanishes, z becomes indeterminate, likewise x and y, and the line lies wholly in the plane.

If the plane $Ax + By + Cz + D = 0$ contains the point (x', y', z') and the line $x = az + \alpha$, $y = bz + \beta$,
$$Ax' + By' + Cz' + D = 0, \quad Aa + Bb + C = 0, \quad A\alpha + B\beta + D = 0.$$

These equations determine the relative values of A, B, C, D, hence the plane is determined.

The plane $Ax + By + Cz + D = 0$ contains the two lines $x = az + \alpha$, $y = bz + \beta$ and $x = a'z + \alpha'$, $y = b'z + \beta'$ when $Aa + Bb + C = 0$, $A\alpha + B\beta + D = 0$, $Aa' + Bb' + C = 0$, $A\alpha' + B\beta' + D = 0$. These four equations are consistent only when $\dfrac{a-a'}{b-b'} = \dfrac{\alpha-\alpha'}{\beta-\beta'}$, that is, when the lines intersect, and then the relative values of A, B, C, D, which determine the plane, are found by solving any three of the four equations.

Problems. — 1. Find angle between planes $10x - 3y + 4z + 12 = 0$, $15x + 11y - 7z + 20 = 0$.

2. Find angle between line $x = 5z + 7$, $y = 3z - 2$, and plane $2x - 15y + 20z + 18 = 0$.

3. Find equation of plane through $(4, -2, 3)$ parallel to $3x - 2y + z - 5 = 0$.

4. Find equation of plane through $(1, 2, -1)$ containing the line $x = 2z - 3$, $y = z + 5$.

5. Find equation of line through $(4, 2, -3)$ perpendicular to $x + 3y - 2z + 4 = 0$.

6. Find equation of plane containing the lines $x = 2z + 1$, $y = 2z + 2$, and $x = z + 5$, $y = 4z - 6$.

7. Find angles which $Ax + By + Cz + D = 0$ makes with the coordinate axes.

8. Find angles which $Ax + By + Cz + D = 0$ makes with the coordinate planes.

9. Show that if two planes are parallel, their traces are parallel.

10. Show that if a line is perpendicular to a plane, the projections of the line are perpendicular to the traces of the plane.

11. Show that $\dfrac{x-x'}{A} = \dfrac{y-y'}{B} = \dfrac{z-z'}{C}$ is perpendicular to
$$Ax + By + Cz + D = 0.$$

12. Show that $(x'-x'')(x-x'')+(y'-y'')(y-y'')+(z'-z'')(z-z'')=0$ is a plane through (x'', y'', z'') perpendicular to the line through (x', y', z') and (x'', y'', z'').

13. Find the equation of the plane tangent to the sphere $x^2+y^2+z^2=R^2$ at the point (x'', y'', z'') of the surface.

14. Find the equation of the plane tangent to the sphere
$$(x-x')^2 + (y-y')^2 + (z-z')^2 = R^2$$
at the point (x'', y'', z'') of the surface.

CHAPTER XIV

CURVED SURFACES

ART. 95.—Cylindrical Surfaces

Let the straight line $x = az + \alpha$, $y = bz + \beta$ move in such a manner that it always intersects the XY-plane in the curve $F(x, y) = 0$, and remains parallel to its first position. The straight line is the generatrix, the curve $F(x, y) = 0$ the directrix of a cylindrical surface. The generatrix pierces the XY-plane in the point (α, β), and therefore $F(\alpha, \beta) = 0$. This is the line equation of the cylindrical surface, for since a and b are constant, to every pair of values of α and β there corresponds one position of the generatrix, and to all pairs of values of α and β satisfying the equation $F(\alpha, \beta) = 0$ there corresponds the generatrix in all positions while generating the cylindrical surface. To obtain the equation of the cylindrical surface in terms of the coordinates of any point (x, y, z) of the surface, substitute in $F(\alpha, \beta) = 0$ the values of α and β obtained from the equations of the generatrix. There results $F(x - az, y - bz) = 0$, the equation of the cylindrical surface whose directrix is $F(x, y) = 0$, generatrix $x = az + \alpha$, $y = bz + \beta$.

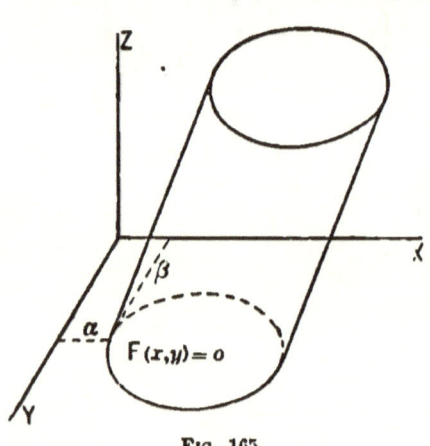

Fig. 165.

CURVED SURFACES

Problems. — 1. Find the equation of the right circular cylinder whose directrix is $x^2 + y^2 = r^2$, and axis the Z-axis.

2. The directrix of a cylinder is a circle in the XY-plane, center at origin. The element of the cylinder in the ZX-plane makes an angle of 45° with the X-axis. Find equation of surface of cylinder.

3. Find general equation of surface of cylinder whose directrix is $\dfrac{x^2}{a^2} + \dfrac{y^2}{b^2} = 1$. What does this equation become when elements are parallel to Z-axis?

4. Find equation of cylindrical surface directrix $y^2 = 10x - x^2$, elements parallel to $x = 2z + 5$, $y = -3z + 5$.

5. Determine locus represented by

$x = a \sin \phi$, $y = a \cos \phi$, $z = c\phi$.

Since $x^2 + y^2 = a^2$, the locus must lie on the cylindrical surface whose axis is the Z-axis, radius of base a. Points corresponding to values of ϕ differing by 2π lie in the same element of the cylindrical surface. The distance between the successive points of intersection of an element of the cylindrical surface with the locus is $2\pi c$. The locus is therefore the thread of a cylindrical screw with distance between threads $2\pi c$. The curve is called the helix.

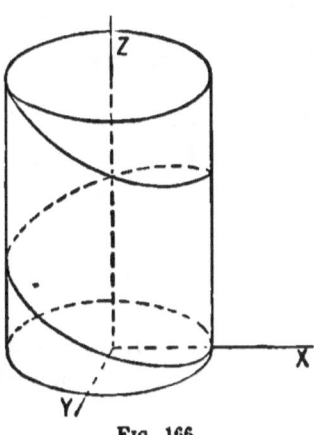

Fig. 166.

Art. 96. — Conical Surfaces

Let the straight line $x = az + \alpha$, $y = bz + \beta$ move in such a manner that it always intersects the XY-plane in the curve $F(x, y) = 0$, and passes through the point (x', y', z'). The straight line generates a conical surface whose vertex is (x', y', z'), directrix $F(x, y) = 0$. The equations of the generatrix are $x - x' = a(z - z')$, $y - y' = b(z - z')$, which may be written $x = az + (x' - az')$, $y = bz + (y' - bz')$. This line pierces the XY-

plane in $(x'-az', y'-bz')$, and therefore $F(x'-az', y'-bz')=0$. This is the line equation of the conical surface, for to every pair of values of a and b there corresponds one position of the generatrix, and to all pairs of values of a and b satisfying the equation $F(x'-az', y'-bz)=0$ there corresponds the generatrix in all positions while generating the conical surface. To obtain the equation of the conical surface in terms of the coordinates of any point (x, y, z) of the surface, substitute in $F(x' - az', y' - bz')= 0$ for a and b their values obtained from the equations of the generatrix. There results $F\left(\dfrac{x'z - xz'}{z - z'}, \dfrac{y'z - yz'}{z - z'}\right)= 0$, the equation of the conical surface whose vertex is (x', y', z'), directrix $F(x, y)= 0$.*

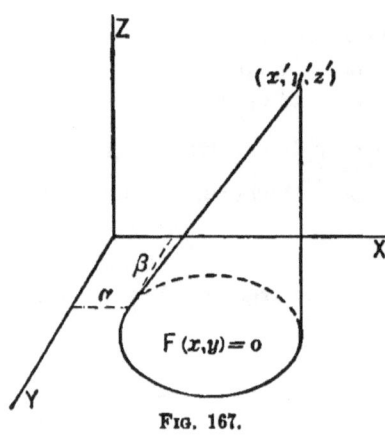

Fig. 167.

Problems. — 1. Find the equation of the surface of the right circular cone whose axis coincides with the Z-axis, vertex at a distance c from the origin.

2. Find the equation of the conical surface directrix $\dfrac{x^2}{9}+\dfrac{y^2}{4}=1$, vertex $(5, 2, 1)$.

3. Find the equation of the conical surface vertex $(0, 0, 10)$, directrix $y^2 = 10x - x^2$.

4. Find the equation of the conical surface vertex $(0, 0, c)$, directrix $\dfrac{x^2}{a^2}+\dfrac{y^2}{b^2}=1$.

5. Find the equation of the conical surface vertex $(0, 0, 10)$, directrix $x^2 + y^2 = 9$.

* Surfaces which may be generated by a straight line are called ruled surfaces.

CURVED SURFACES

Art. 97. — Surfaces of Revolution

Let MN be any line in the ZX-plane. When MN revolves about the Z-axis, every point P of MN describes the circumference of a circle with its center on the Z-axis and which is projected on the XY-plane in an equal circle. The equation of the circle referred to a pair of axes through its center parallel to the axes X and Y is

$$x^2 + y^2 = r^2.$$

This is also the equation of the projection of the circle on the XY-plane. The radius r is a function of z which is given by

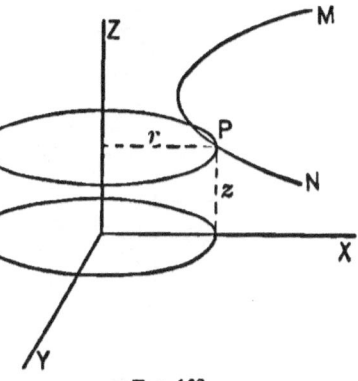

Fig. 168.

the equation of the generatrix $r = F(z)$. Hence the equation of the surface of revolution is obtained by eliminating r from the equations $x^2 + y^2 = r^2$ and $r = F(z)$.

Problems. — 1. Find equation of surface of sphere, center at origin, radius R. This sphere is generated by the revolution about the Z-axis of a circle whose equation is $r^2 + z^2 = R^2$. Eliminate r from this equation and $x^2 + y^2 = r^2$, and the equation of the sphere is found to be

$$x^2 + y^2 + z^2 = R^2.$$

2. Find equation of right circular cylinder whose axis is the Z-axis.

3. Find equation of right circular cone whose axis is Z-axis, vertex $(0, 0, c)$.

4. Find equation of right circular cone whose axis is Z-axis, vertex $(0, 0, 0)$.

5. Find equation of surface generated by revolution of ellipse about its major axis. This is the prolate spheroid.

6. Find equation of surface generated by revolution of ellipse about its minor axis. This is the oblate spheroid.

7. Find equation of surface generated by revolution of hyperbola about its conjugate axis. This is the hyperboloid of revolution of one sheet.

8. Find equation of surface generated by revolution of hyperbola about its transverse axis. This is the hyperboloid of revolution of two sheets.

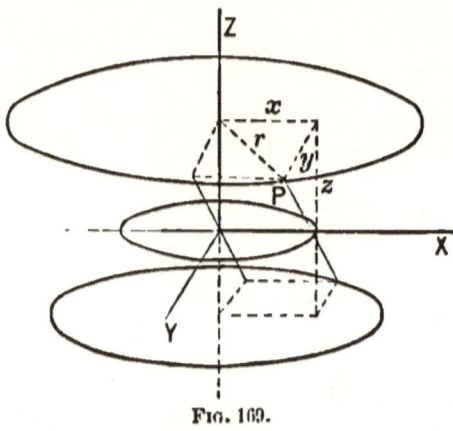

Fig. 169.

9. Let PP'' be perpendicular to the X-axis, but not in the ZX-plane. Suppose PP'' to revolve about the Z-axis. The equation of the surface generated is to be found.

The equations of the projections of PP'' on the planes ZX and ZY are $x = a$, $y = bz$. The point P describes the circumference of a circle whose equation is $x^2 + y^2 = r^2$. The value of r depends on z, and from the figure $r^2 = a^2 + b^2 z^2$. Hence the equation of the surface generated is $x^2 + y^2 = b^2 z^2 + a^2$. The surface is therefore an hyperboloid of revolution of one sheet.

Art. 98. — The Ellipsoid

In the XY-plane there is the fixed ellipse $\dfrac{x^2}{a^2} + \dfrac{y^2}{b^2} = 1$, in the ZX-plane the fixed ellipse $\dfrac{x^2}{a^2} + \dfrac{z^2}{c^2} = 1$. The figure generated

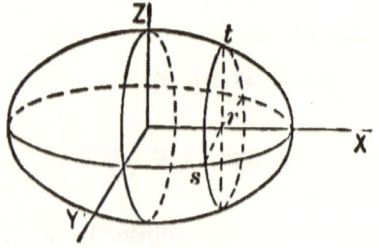

Fig. 170.

by the ellipse which moves with its center on the X-axis, the plane of the ellipse perpendicular to the X-axis, the axes of the ellipse in any position the intersections of the plane of the ellipse with the fixed ellipses, is called the ellipsoid. The equation of the ellipse is $\dfrac{y^2}{\overline{rs}^2} + \dfrac{z^2}{\overline{rt}^2} = 1$. From the equations of the fixed ellipses

$\frac{x^2}{a^2}+\frac{\overline{rs}^2}{b^2}=1$, $\frac{x^2}{a^2}+\frac{\overline{rt}^2}{c^2}=1$, whence $\overline{rs}^2=b^2\left(1-\frac{x^2}{a^2}\right)$, $\overline{rt}^2=c^2\left(1-\frac{x^2}{a^2}\right)$.
Hence the equation of the generating ellipse in any position, that is, the equation of the ellipsoid, is $\frac{x^2}{a^2}+\frac{y^2}{b^2}+\frac{z^2}{c^2}=1$. When a, b, c are unequal, the figure is an ellipsoid with unequal axes; when two of the axes are equal, the figure is an ellipsoid of revolution, or spheroid; when the three axes are equal, the ellipsoid becomes the sphere.

Art. 99. — The Hyperboloids

In the ZX-plane there is the fixed hyperbola $\frac{x^2}{a^2}-\frac{z^2}{c^2}=1$, in the ZY-plane the fixed hyperbola $\frac{y^2}{b^2}-\frac{z^2}{c^2}=1$. The figure generated by the ellipse which moves with its center on the Z-axis, the plane of the ellipse perpendicular to the Z-axis, the axes of the ellipse in any position the intersections of the plane of the ellipse with the fixed hyperbolas, is called the hyperboloid of one sheet. The equation of the ellipse is $\frac{x^2}{\overline{rs}^2}+\frac{y^2}{\overline{rt}^2}=1$. From the equations of the fixed hyperbolas

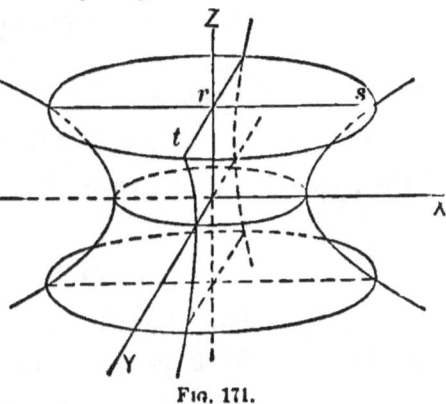

Fig. 171.

$\frac{\overline{rs}^2}{a^2}-\frac{z^2}{c^2}=1$, $\frac{\overline{rt}^2}{b^2}-\frac{z^2}{c^2}=1$, whence $\overline{rs}^2=a^2\left(1+\frac{z^2}{c^2}\right)$, $\overline{rt}^2=b^2\left(1+\frac{z^2}{c^2}\right)$.
Hence the equation of the generating ellipse in any position, that is, the equation of the hyperboloid of one sheet, is $\frac{x^2}{a^2}+\frac{y^2}{b^2}-\frac{z^2}{c^2}=1$.

In the ZX-plane there is the fixed hyperbola $\frac{x^2}{a^2}-\frac{z^2}{c^2}=1$,

in the XY-plane the fixed hyperbola $\dfrac{x^2}{a^2} - \dfrac{y^2}{b^2} = 1$. The figure generated by the ellipse which moves with its center on the

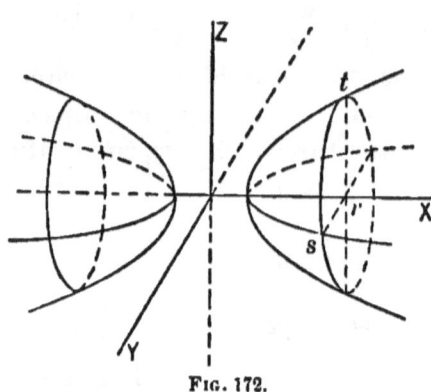

Fig. 172.

X-axis, the plane of the ellipse perpendicular to the X-axis, the axes of the ellipse in any position the intersections of the plane of the ellipse with the fixed hyperbolas, is called the hyperboloid of two sheets. The equation of the ellipse is $\dfrac{y^2}{\overline{rs}^2} + \dfrac{z^2}{\overline{rt}^2} = 1$. From the equations of the fixed hyperbolas $\dfrac{x^2}{a^2} - \dfrac{\overline{rs}^2}{b^2} = 1$, $\dfrac{x^2}{a^2} - \dfrac{\overline{rt}^2}{c^2} = 1$, whence $\overline{rs}^2 = b^2\left(\dfrac{x^2}{a^2} - 1\right)$, $\overline{rt}^2 = c^2\left(\dfrac{x^2}{a^2} - 1\right)$. Hence the equation of the generating ellipse in any position, that is, the equation of the hyperboloid of two sheets, is $\dfrac{x^2}{a^2} - \dfrac{y^2}{b^2} - \dfrac{z^2}{c^2} = 1$.

Art. 100.—The Paraboloids

In the XY-plane there is the fixed parabola $y^2 = 2bx$, in the ZX-plane the fixed parabola $z^2 = 2cx$. The figure generated

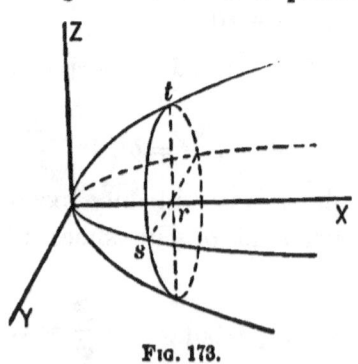

Fig. 173.

by an ellipse which moves with its center on the X-axis, its plane perpendicular to the X-axis, the axes of the ellipse in any position the intersections of the plane of the ellipse with the fixed parabola, is called the elliptical paraboloid. The equation of the ellipse is

$$\dfrac{y^2}{\overline{rs}^2} + \dfrac{z^2}{\overline{rt}^2} = 1.$$

From the equations of the fixed parabolas $\overline{rs}^2 = 2\,bx$, $\overline{rt}^2 = 2\,cx$. Hence the equation of the generating ellipse in any position, that is, the equation of the elliptical paraboloid, is $\dfrac{y^2}{b} + \dfrac{z^2}{c} = 2\,x$.

In the ZX-plane there is the fixed parabola $z^2 = 2\,cx$, in the XY-plane the fixed parabola $y^2 = -2\,bx$. The figure generated by an hyperbola which moves with its center on the X-axis, the plane of the hyperbola perpendicular to the X-axis, the axes of the hyperbola the intersections of the plane of the hyperbola with the fixed parabolas, is called the hyperbolic paraboloid. The equation of the hyperbola is

$$\dfrac{z^2}{\overline{rs}^2} - \dfrac{y^2}{\overline{rt}^2} = 1.$$

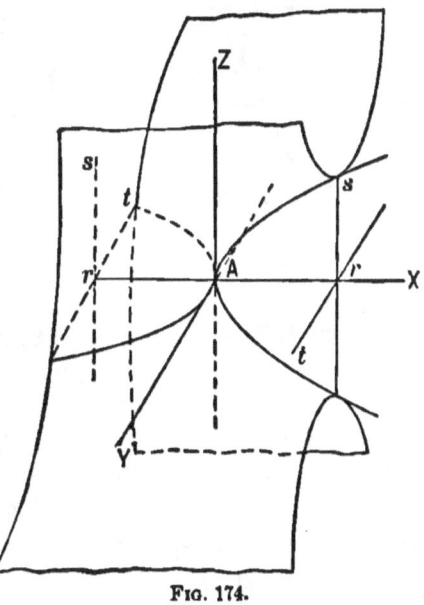

Fig. 174.

From the equations of the fixed parabolas $\overline{rs}^2 = 2\,cx$, $\overline{rt}^2 = -2\,bx$. Hence the equation of the generating hyperbola in any position, that is, the equation of the hyperbolic paraboloid, is $\dfrac{z^2}{c} - \dfrac{y^2}{b} = 2\,x$.

Art. 101. — The Conoid

The center of an ellipse moves in a straight line perpendicular to the plane of the ellipse. The major axis is constant for all positions of the ellipse, the minor axis diminishes directly as the distance the ellipse has moved, becoming zero when the ellipse has moved the distance c. The figure generated is called the conoid with elliptical base. The equation of the ellipse is

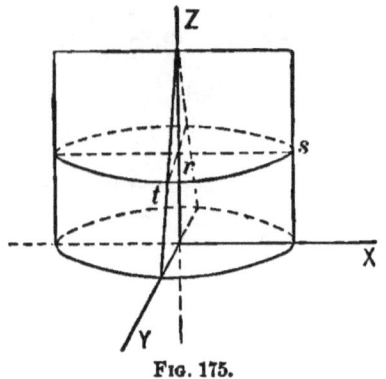

Fig. 175.

$$\frac{x^2}{rs^2}+\frac{y^2}{rt^2}=1,$$

where $rs = a$, and, from similar triangles, $\frac{rt}{b}=\frac{c-z}{c}$, whence $rt = \frac{b}{c}(c-z)$. The equation of the generating ellipse in any position, that is, the equation of the conoid, is $\frac{c^2 y^2}{b^2(c-z)^2}+\frac{x^2}{a^2}=1$.

Art. 102. — Surfaces represented by Equations in Three Variables

An equation $\phi(x, y, z) = 0$, when interpreted in rectangular space coordinates, represents some surface. For when z is a continuous function of x and y, if (x, y) takes consecutive positions in the XY-plane, the point (x, y, z) takes consecutive positions in space. Hence the geometric representation of the function $\phi(x, y, z) = 0$ is the surface into which this function transforms the XY-plane. To determine the form and dimensions of the surface represented by a given equation, the intersections of this surface by planes parallel to the coordinate planes are studied.

Problems. — Determine the form and dimensions of the surfaces represented by the following equations.

1. $\frac{x^2}{9}+\frac{y^2}{4}+z^2=1$. The equation of the projection on the XY-plane of the intersection of the surface represented by this equation and a plane $z = c$ parallel to the XY-plane is $\frac{x^2}{9}+\frac{y^2}{4}=1-c^2$. This equation repre-

sents an ellipse whose dimensions are greatest when $c = 0$, diminish as the numerical value of c increases to 1, and are zero when $c = \pm 1$. The ellipse is imaginary when c is numerically greater than 1.

The equation of the projection on the ZX-plane of the intersection of the surface by a plane $y = b$ parallel to the ZX-plane is $\dfrac{x^2}{9} + z^2 = 1 - \dfrac{b^2}{4}$, which represents an ellipse whose dimensions are greatest when $b = 0$, diminish as b increases numerically to 2, are zero when $b = \pm 2$, and become imaginary when b is numerically greater than 2.

The equation of the projection on the YZ-plane of the intersection of the surface by a plane $x = a$ parallel to the YZ-plane is $\dfrac{y^2}{4} + z^2 = 1 - \dfrac{a^2}{9}$,

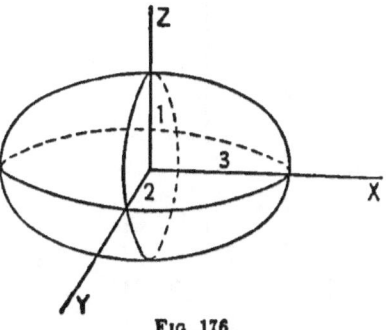

Fig. 176.

which represents an ellipse whose dimensions are greatest when $a = 0$, diminish as a increases numerically to 3, are zero when $a = \pm 3$, and become imaginary when a is numerically greater than 3.

The sections of the surface made by planes parallel to the coordinate planes are all ellipses, the surface is closed and limited by the faces of the rectangular parallelopiped whose faces are $x = \pm 3$, $y = \pm 2$, $z = \pm 1$.

From the equation it is seen that the origin is a center of symmetry, the coordinate axes are axes of symmetry, the coordinate planes are planes of symmetry of the surface. The figure is the ellipsoid with axes 3, 2, 1.

2. $x^2 + y^2 - z^2 = 0$.
3. $x^2 + y^2 + z^2 - 10x = 0$.
4. $y^2 + z^2 - 10x = 0$.
5. $x^2 + y^2 - z^2 = 1$.
6. $x^2 - y^2 - z^2 = 1$.
7. $z^2 - 2x + 3y = 5$.
8. $z^2 + 2x + 4z = 7$.
9. $x^2 + y^2 + 10z + 15 = 0$.
10. $\dfrac{2y^2}{(2-z)^2} + x^2 = 1$.
11. $\dfrac{x^2}{a^2} + \dfrac{y^2}{b^2} - \dfrac{z^2}{c^2} = 0$.
12. Show that the conoid is a ruled surface.

CHAPTER XV

SECOND DEGREE EQUATION IN THREE VARIABLES

Art. 103. — Transformation of Coordinates

Take the point (a, b, c) referred to the axes X, Y, Z as the origin of a set of axes X', Y', Z' parallel to X, Y, Z respectively. Let (x, y, z), (x', y', z') represent the same point referred to the two sets of axes. From the figure $x = x' + a$, $y = y' + b$, $z = z' + c$.

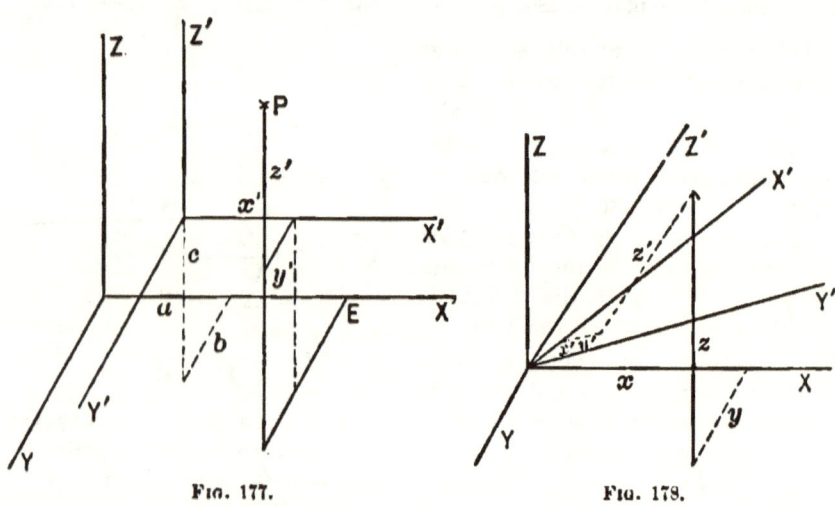

Fig. 177.   Fig. 178.

Let X, Y, Z be a set of rectangular axes, X', Y', Z' any set of rectilinear axes with the same origin. Denote the angles made by X' with X, Y, Z by α, β, γ respectively, the angles made by Y' with X, Y, Z by α', β', γ', the angles made by Z' with X, Y, Z by α'', β'', γ''. If (x, y, z) and (x', y', z') represent the same point P, x is the projection of the broken line

SECOND DEGREE EQUATION

$(x' + y' + z')$ on the X-axis, y the projection of this broken line on the Y-axis, z the projection of this broken line on the Z-axis. Hence

$$x = x' \cos \alpha + y' \cos \alpha' + z' \cos \alpha'',$$
$$y = x' \cos \beta + y' \cos \beta' + z' \cos \beta'',$$
$$z = x' \cos \gamma + y' \cos \gamma' + z' \cos \gamma''.$$

Since X, Y, Z are rectangular axes,

$$\cos^2 \alpha + \cos^2 \beta + \cos^2 \gamma = 1,$$
$$\cos^2 \alpha' + \cos^2 \beta' + \cos^2 \gamma' = 1,$$
$$\cos^2 \alpha'' + \cos^2 \beta'' + \cos^2 \gamma'' = 1.$$

If X', Y', Z' are also rectangular,

$$\cos \alpha \cos \alpha' + \cos \beta \cos \beta' + \cos \gamma \cos \gamma' = 0,$$
$$\cos \alpha \cos \alpha'' + \cos \beta \cos \beta'' + \cos \gamma \cos \gamma'' = 0,$$
$$\cos \alpha' \cos \alpha'' + \cos \beta' \cos \beta'' + \cos \gamma' \cos \gamma'' = 0.$$

Problems.—1. Transform $x^2 + y^2 + z^2 = 25$ to parallel axes, origin $(-5, 0, 0)$.

2. Transform $x^2 + y^2 + z^2 = 25$ to parallel axes, origin $(-5, -5, -5)$.

3. Transform $\dfrac{x^2}{a^2} + \dfrac{y^2}{b^2} + \dfrac{z^2}{c^2} = 1$ to parallel axes, origin $(-a, 0, 0)$.

4. Show that the first degree equation in three variables interpreted in oblique coordinates represents a plane.

5. Show that the equation of an elliptic cone, vertex at origin, and axis the Z-axis, is $\dfrac{x^2}{a^2} + \dfrac{y^2}{b^2} - \dfrac{z^2}{c^2} = 0$.

6. Derive the formulas for transformation from one rectangular system to another rectangular system, the Z'-axis coinciding with the Z-axis, the X'-axis making an angle θ with the X-axis.

ART. 104.—PLANE SECTION OF QUADRIC

Surfaces represented by the second degree equation in three variables

$$Ax^2 + By^2 + Cz^2 + 2Dxy + 2Exz + 2Fyz$$
$$+ 2Gx + 2Hy + 2Kz + L = 0 \quad (1)$$

are known by the general name of quadrics.

To find the intersection of the surface represented by this equation by any plane transform to a set of axes parallel to the original set, having some point (a, b, c) in the cutting plane as origin. The transformation formulas are

$$x = x' + a, \qquad y = y' + b, \qquad z = z' + c,$$

and the transformed equation is

$$Ax'^2 + By'^2 + Cz'^2 + 2D'x'y' + 2E'x'z'$$
$$+ 2F'y'z' + 2G'x' + 2H'y' + 2K'z' + L' = 0, \quad (2)$$

where
$$G' = Aa + Db + Ec + G,$$
$$H' = Bb + Da + Fc + H, \qquad K' = Cc + Ea + Fb + K,$$
$$L' = Aa^2 + Bb^2 + Cc^2 + 2Dab + 2Eac$$
$$+ 2Fbc + 2Ga + 2Hb + 2Kc + L.$$

Now turn the axes X', Y', Z' about the origin until the $X'Y'$-plane coincides with the cutting plane. This is accomplished by the transformation formulas

$$x' = x_1 \cos \alpha + y_1 \cos \alpha' + z_1 \cos \alpha'',$$
$$y' = x_1 \cos \beta + y_1 \cos \beta' + z_1 \cos \beta'',$$
$$z' = x_1 \cos \gamma + y_1 \cos \gamma' + z_1 \cos \gamma''.$$

These formulas are linear, hence the equation of the quadric in terms of (x_1, y_1, z_1) is of the form

$$A_1 x_1^2 + B_1 y_1^2 + C_1 z_1^2 + 2D_1 x_1 y_1 + 2E_1 x_1 z_1$$
$$+ 2F_1 y_1 z_1 + 2G_1 x_1 + 2H_1 y_1 + 2K_1 z_1 + L_1 = 0. \quad (3)$$

Since the plane of the section is the $X_1 Y_1$-plane, the equation of the intersection referred to rectangular axes in its own plane is $A_1 x_1^2 + B_1 y_1^2 + 2 D_1 x_1 y_1 + 2 G_1 x_1 + 2 H_1 y_1 + L_1 = 0$, which represents a conic section. Hence every plane section of a quadric is a conic section. For this reason quadrics are also called conicoids.

Art. 105. — Center of Quadric

The surface represented by equation (2) is symmetrical with respect to the origin (a, b, c) when the coefficients of x', y', z' are zero, for then if (x', y', z') is a point of the surface,
$$(-x', -y', -z')$$
is also a point of the surface. Hence the center of the quadric (1) is found by solving the equations

$$Aa + Db + Ec + G = 0, \qquad Bb + Da + Fc + H = 0,$$
and
$$Cc + Ea + Fb + K = 0.$$

Problems. — 1. Find the center of the quadric represented by
$$x^2 + y^2 + 4z^2 - 8xz + 6y = 0.$$

2. Find the center of the quadric represented by
$$x^2 - y^2 + z^2 - 10x + 8z + 15 = 0.$$

Art. 106. — Tangent Plane to Quadric

Let (x_0, y_0, z_0) be any point of the quadric (1). The equations $x = x_0 + d\cos\alpha$, $y = y_0 + d\cos\beta$, $z = z_0 + d\cos\gamma$ represent all straight lines through (x_0, y_0, z_0). By substituting in (1)

$+ A x_0^2$	$+ 2A\cos\alpha \cdot x_0$	$d + A\cos^2\alpha$	$d^2 = 0$
$+ B y_0^2$	$+ 2B\cos\beta \cdot y_0$	$+ B\cos^2\beta$	
$+ C z_0^2$	$+ 2C\cos\gamma \cdot z_0$	$+ C\cos^2\gamma$	
$+ 2D x_0 y_0$	$+ 2D\cos\alpha \cdot y_0$	$+ 2D\cos\alpha\cos\beta$	
$+ 2E x_0 z_0$	$+ 2D\cos\beta \cdot x_0$	$+ 2E\cos\alpha\cos\gamma$	
$+ 2F y_0 z_0$	$+ 2E\cos\gamma \cdot x_0$	$+ 2F\cos\beta\cos\gamma$	
$+ 2G x_0$	$+ 2E\cos\alpha \cdot z_0$		
$+ 2H y_0$	$+ 2F\cos\gamma \cdot y_0$		
$+ 2K z_0$	$+ 2F\cos\beta \cdot z_0$		
$+ L$	$+ 2G\cos\alpha$		
	$+ 2H\cos\beta$		
	$+ 2K\cos\gamma$		

an equation is found which determines the two values of d corresponding to the points of intersection of straight line and quadric. Since the point (x_0, y_0, z_0) lies in the quadric, the term of this equation independent of d vanishes. If the coefficient of the first power of d also vanishes, the equation has two roots equal to zero; that is, every straight line through the point (x_0, y_0, z_0), and whose direction cosines satisfy the equation

$$A \cos \alpha \cdot x_0 + B \cos \beta \cdot y_0 + C \cos \gamma \cdot z_0 + D \cos \alpha \cdot y_0 + D \cos \beta \cdot x_0$$
$$+ E \cos \gamma \cdot x_0 + E \cos \alpha \cdot z_0 + F \cos \gamma \cdot y_0 + F \cos \beta \cdot z_0$$
$$+ G \cos \alpha + H \cos \beta + K \cos \gamma = 0$$

is tangent to the quadric. To determine the surface represented by this equation multiply by d and substitute $x - x_0$ for $d \cos \alpha$, $y - y_0$ for $d \cos \beta$, $z - z_0$ for $d \cos \gamma$. There results the equation

$$Axx_0 + Byy_0 + Czz_0 + D(xy_0 + x_0y) + E(xz_0 + x_0z)$$
$$+ F(yz_0 + y_0z) + G(x + x_0) + H(y + y_0) + K(z + z_0) + L = 0,$$

which, since it is of the first degree in (x, y, z) represents a plane. This plane, containing all the straight lines tangent to the quadric at (x_0, y_0, z_0) is tangent to the quadric at (x_0, y_0, z_0). Notice that the equation of the plane tangent to the quadric at (x_0, y_0, z_0) is obtained by substituting in the equation of the quadric xx_0 for x^2, yy_0 for y^2, zz_0 for z^2, $xy_0 + x_0y$ for $2xy$, $xz_0 + x_0z$ for $2xz$, $yz_0 + y_0z$ for $2yz$, $x + x_0$ for $2x$, $y + y_0$ for $2y$, $z + z_0$ for $2z$.

Let (x', y', z') be any point in space, (x_0, y_0, z_0) the point of contact with the quadric (1) of any plane through (x', y', z') tangent to the quadric. Then (x_0, y_0, z_0), (x', y', z') must satisfy the equation

$$Ax'x_0 + By'y_0 + Cz'z_0 + D(x'y_0 + y'x_0) + E(z'x_0 + x'z_0)$$
$$+ F(z'y_0 + y'z_0) + G(x' + x_0) + H(y' + y_0) + K(z' + z_0) + L = 0.$$

Hence the points of contact (x_0, y_0, z_0) must lie in a plane, and the locus of the points of contact is a conic section.

Problems. — 1. Write the equation of the plane tangent to
$$x^2 + y^2 + z^2 = R^2 \text{ at } (x_0, y_0, z_0).$$

2. Write the equation of the plane tangent to
$$x^2 + y^2 + z^2 - 10x + 25 = 0 \text{ at } (5, 0, 0).$$

3. Write the equation of the plane tangent to
$$\frac{x^2}{a^2} + \frac{y^2}{b^2} + \frac{z^2}{c^2} = 1 \text{ at } (x_0, y_0, z_0).$$

4. Write the equation of the plane tangent to
$$\frac{y^2}{b} + \frac{z^2}{c} = 2x \text{ at } (x_0, y_0, z_0).$$

5. Find equations of projections on planes ZX and ZY of locus of points of contact of planes tangent to $x^2 + y^2 + z^2 = 25$ through $(7, -10, 6)$.

6. Find equation of normal to $\frac{x^2}{a^2} + \frac{y^2}{b^2} + \frac{z^2}{c^2} = 1$ at (x', y', z'). The normal to a surface at any point is the line through that point perpendicular to the tangent plane at that point.

7. Find the angle between the normal to $\frac{x^2}{a^2} + \frac{y^2}{b^2} + \frac{z^2}{c^2} = 1$ at (x', y', z') and the line joining (x', y', z') and the center of the ellipsoid.

ART. 107. — REDUCTION OF GENERAL EQUATION OF QUADRIC

To determine the form and dimensions of the surfaces represented by the general second degree equation in three variables when interpreted in rectangular space coordinates it is desirable first to simplify the equation. This simplification is effected by changing the position of the origin and the direction of the axes.

The change of direction of rectangular axes is effected by the formulas
$$x = x' \cos \alpha + y' \cos \alpha' + z' \cos \alpha'',$$
$$y = x' \cos \beta + y' \cos \beta' + z' \cos \beta'',$$
$$z = x' \cos \gamma + y' \cos \gamma' + z' \cos \gamma'',$$

where the nine cosines are subject to the six conditions

$$\cos^2 \alpha + \cos^2 \beta + \cos^2 \gamma = 1,$$
$$\cos^2 \alpha' + \cos^2 \beta' + \cos^2 \gamma' = 1,$$
$$\cos^2 \alpha'' + \cos^2 \beta'' + \cos^2 \gamma'' = 1,$$
$$\cos \alpha \cos \alpha' + \cos \beta \cos \beta' + \cos \gamma \cos \gamma' = 0,$$
$$\cos \alpha \cos \alpha'' + \cos \beta \cos \beta'' + \cos \gamma \cos \gamma'' = 0,$$
$$\cos \alpha' \cos \alpha'' + \cos \beta' \cos \beta'' + \cos \gamma' \cos \gamma'' = 0.$$

Three arbitrary conditions may therefore be imposed on the nine cosines.

Substituting for x, y, z in

$$Ax^2 + By^2 + Cz^2 + 2\,Dxy + 2\,Exz + 2\,Fyz + 2\,Gx$$
$$+ 2\,Hy + 2\,Kz + L = 0,$$

there results

$$Ax'^2 + By'^2 + Cz'^2 + 2\,D'x'y' + 2\,E'x'z' + 2\,F'y'z' + 2\,G'x'$$
$$+ 2\,H'y' + 2\,K'z' + L' = 0,$$

when D', E', F' are functions of the nine cosines. Equate D', E', F' to zero and determine the directions of the rectangular coordinates in space in accordance with these equations. This transformation is always possible, hence

$$Ax^2 + By^2 + Cz^2 + 2\,G'x + 2\,H'y + 2\,K'z + L' = 0$$

interpreted in rectangular coordinates represents all quadric surfaces.

Now transform to parallel axes with the origin at (a, b, c). The transformation formulas are

$$x = a + x', \quad y = b + y', \quad z = c + z'$$

and the transformed equation

$$Ax'^2 + By'^2 + Cz'^2 + 2(Aa + G')x' + 2(Bb + H')y' + 2(Cc + K')z'$$
$$+ (Aa^2 + Bb^2 + Cc^2 + 2\,G'a + 2\,H'b + 2\,K'c + L') = 0.$$

Take advantage of the three arbitrary constants a, b, c to cause the vanishing of the coefficients of x', y', z'. This gives

$$a = -\frac{G'}{A}, \quad b = -\frac{H'}{B}, \quad c = -\frac{K'}{C},$$

values which are admissible when $A \neq 0, B \neq 0, C \neq 0$. The resulting equation is of the form $Lx^2 + My^2 + Nz^2 = P$.

When $A \neq 0, B \neq 0, C = 0$, the transformation

$$x = a + x', \quad y = b + y', \quad z = c + z'$$

gives
$$Ax'^2 + By'^2 + 2(Aa + G')x' + 2(Bb + H')y' + K'z'$$
$$+ (Aa^2 + Bb^2 + 2G'a + 2H'b + 2K'c + L') = 0.$$

Equating to zero the coefficients of x', y' and the absolute term, the values found for a, b, c are finite when $A \neq 0, B \neq 0, K' \neq 0$. The resulting equation is of the form $Lx^2 + My^2 + N'z = 0$.

When $A \neq 0, B \neq 0, C = 0, K' = 0$, the equation takes the form $Lx^2 + My^2 + L'x + M'y + P = 0$.

When $A \neq 0, B = 0, C = 0$, the equation takes the form $Mx^2 + M'x + N'y + P = 0$.

When $A = 0, B = 0, C = 0$, the equation is no longer of the second degree.

Since x, y, z are similarly involved in

$$Ax^2 + By^2 + Cz^2 + 2G'x + 2H'y + 2K'z + L' = 0,$$

the vanishing of A and G' or of B and H' would lead to equations of the same form as the vanishing of C and K'.

Collecting results it is seen that the following equations interpreted in rectangular coordinates represent all quadric surfaces —

$A \neq 0, B \neq 0, C \neq 0,$	$Lx^2 + My^2 + Nz^2 = P$	I
$A \neq 0, B \neq 0, C = 0, K' \neq 0$	$Lx^2 + My^2 + N'z = 0$	II
$A \neq 0, B \neq 0, C = 0, K' = 0$	$Lx^2 + My^2 + M'y + L'x + P = 0$	III
$A \neq 0, B = 0, C = 0$	$Lx^2 + L'x + M'y + N'z + P = 0$	

These equations are known as equations of the first, second, and third class.

Art. 108. — Surfaces of the First Class

The equation of the first class may take the forms

(a) $Lx^2 + My^2 + Nz^2 = P$, (b) $Lx^2 + My^2 - Nz^2 = P$,
(c) $Lx^2 + My^2 - Nz^2 = -P$,

or similar forms with the coefficients of x^2 and z^2 or of y^2 and z^2 positive.

(a) The intersections of planes parallel to the coordinate planes with $Lx^2 + My^2 + Nz^2 = P$ are for

$$x = x', \quad My^2 + Nz^2 = P - Lx'^2,$$

an ellipse whose dimensions are greatest when $x' = 0$, diminish as x' increases numerically, are zero for $x' = \pm\sqrt{\dfrac{P}{L}}$, imaginary when x' is numerically greater than $\sqrt{\dfrac{P}{L}}$;

for $\quad y = y', \; Lx^2 + Nz^2 = P - My'^2,$

an ellipse whose dimensions are greatest when $y' = 0$, diminish as y' increases numerically, are zero for $y' = \pm\sqrt{\dfrac{P}{M}}$, imaginary when y' is numerically greater than $\sqrt{\dfrac{P}{M}}$;

for $\quad z = z', \; Lx^2 + My^2 = P - Nz'^2,$

an ellipse whose dimensions are greatest for $z' = 0$, diminish as z' increases numerically, are zero when $z' = \pm\sqrt{\dfrac{P}{N}}$, imaginary when z' is numerically greater than $\sqrt{\dfrac{P}{N}}$.

Calling the semi-diameter on the X-axis a, on the Y-axis b, on the Z-axis c, the equation becomes $\dfrac{x^2}{a^2} + \dfrac{y^2}{b^2} + \dfrac{z^2}{c^2} = 1$, the ellipsoid.

The figure represented by $Lx^2 + My^2 + Nz^2 = -P$ is imaginary. The equation $Lx^2 + My^2 + Nz^2 = 0$ represents the origin.

(b) $Lx^2 + My^2 - Nz^2 = P$. The intersections are for $x = x'$,

$My^2 - Nz^2 = P - Lx'^2$, an hyperbola whose real axis is parallel to the Y-axis when $-\sqrt{\frac{P}{L}} < x' < +\sqrt{\frac{P}{L}}$, parallel to the Z-axis when x' is numerically greater than $\sqrt{\frac{P}{L}}$, and which becomes two straight lines when $x' = \pm \sqrt{\frac{P}{L}}$;

for $\quad\quad\quad y = y', \ Lx^2 - Nz^2 = P - My'^2$,

an hyperbola whose real axis is parallel to the X-axis when

$$\sqrt{\frac{P}{M}} < y' < +\sqrt{\frac{P}{M}},$$

parallel to the Z-axis when y' is numerically greater than $\sqrt{\frac{P}{M}}$, and which becomes two straight lines when $y' = \pm \sqrt{\frac{P}{M}}$;

for $\quad\quad\quad z = z', \ Lx^2 + My^2 = P + Nz'^2$,

an ellipse, always real, whose dimensions are least when $z' = 0$, and increase indefinitely when z' increases indefinitely in numerical value. Calling the intercepts of this surface on the X-axis a, on the Y-axis b, on the Z-axis $c\sqrt{-1}$, the equation becomes $\frac{x^2}{a^2} + \frac{y^2}{b^2} - \frac{z^2}{c^2} = 1$, the hyperboloid of one sheet.

(c) $\quad\quad\quad Lx^2 + My^2 - Nz^2 = -P$.

The intersections are

for $\quad\quad\quad x = x', \ My^2 - Nz^2 = -P - Lx'^2$,

an hyperbola with its real axis parallel to the Z-axis, dimensions least when $x' = 0$, increasing indefinitely with the numerical value of x';

for $\quad\quad\quad y = y', \ Lx^2 - Nz^2 = -P - My'^2$,

an hyperbola with its real axis parallel to the Z-axis, dimensions least when $y' = 0$, increasing indefinitely with the numerical value of y';

for $\quad\quad\quad z = z', \ Lx^2 + My^2 = Lz'^2 - P$,

an ellipse, imaginary when $-\sqrt{\frac{P}{L}} < z' < +\sqrt{\frac{P}{L}}$, dimensions zero for $z' = \pm\sqrt{\frac{P}{L}}$, increasing indefinitely with the numerical value of z'.

Calling the intercepts of this surface on the axes X, Y, Z respectively, $a\sqrt{-1}, b\sqrt{-1}, c$, the equation becomes

$$\frac{x^2}{a^2} + \frac{y^2}{b^2} - \frac{z^2}{c^2} = -1,$$

the hyperboloid of two sheets.

The surfaces of the first class are ellipsoids and hyperboloids.

Art. 109. — Surfaces of the Second Class

The equation of the second class may take the forms

(*a*) $Lx^2 + My^2 \pm N'z = 0$, (*b*) $Lx^2 - My^2 \pm N'z = 0$.

(*a*) $Lx^2 + My^2 = N'z$. The intersections are

for $x = x'$, $My^2 = N'z - Lx'^2$,

a parabola whose parameter is constant, axis parallel to Z-axis, and whose vertex continually recedes from the origin;

for $y = y'$, $Lx^2 = N'z - My'^2$,

a parabola whose parameter is constant, axis parallel to Z-axis, and whose vertex continually recedes from the origin;

for $z = z'$, $Lx^2 + My^2 = N'z'$,

an ellipse whose dimensions are zero for $z' = 0$ and increase indefinitely as z' increases from 0 to $+\infty$, but are imaginary for $z' < 0$.

This surface is the elliptic paraboloid. The equation $Lx^2 + My^2 = -N'z$ represents an elliptic paraboloid real for negative values of z.

(*b*) $Lx^2 - My^2 = N'z$. The intersections are

for $x = x'$, $My^2 = Lx'^2 - N'z$,

a parabola of constant parameter whose axis is parallel to

the Z-axis and whose vertex recedes from the origin as x' increases numerically;

for $\quad\quad y = y', \ Lx^2 = N'z + My'^2,$

a parabola of constant parameter whose axis is parallel to the Z-axis and whose vertex recedes from the origin as y' increases numerically;

for $\quad\quad z = z', \ Lx^2 - My^2 = N'z',$

an hyperbola whose real axis is parallel to the X-axis when $z > 0$, parallel to the Y-axis when $z' < 0$, and which becomes two straight lines when $z' = 0$.

The surface is the hyperbolic paraboloid.

The surfaces of the second class are paraboloids.

Art. 110. — Surfaces of the Third Class

The equation $Lx^2 + My^2 + L'x + M'y + P = 0$ does not contain z and therefore represents a cylindrical surface whose elements are parallel to the Z-axis. The directrix in the XY-plane is an ellipse when L and M have like signs, an hyperbola when L and M have unlike signs.

The surface represented by the equation

$$Lx^2 + L'x + M'y + N'z + P = 0$$

is intersected by the XY-plane in the parabola
$Lx^2 + L'x + M'y + P = 0,$
by the ZX-plane in the parabola
$Lx^2 + L'x + N'z + P = 0,$

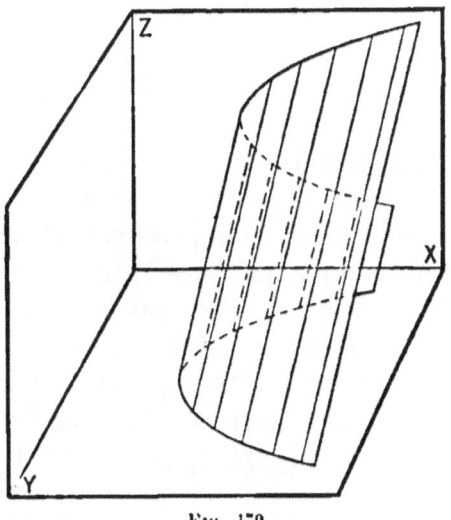

Fig. 179.

by planes $x = x'$ parallel to the YZ-plane in parallel straight lines

$$N'y + L'z + Mx'^2 + M'x' + P = 0.$$

Hence the surface is a parabolic cylinder with elements parallel to the ZY-plane.

The surfaces of the third class are cylindrical surfaces with elliptic, hyperbolic, or parabolic bases.

It is now seen that the second degree equation in three variables represents ellipsoids, hyperboloids, paraboloids, and cylindrical surfaces with conic sections as bases. Conical surfaces are varieties of hyperboloids.

Art. 111. — Quadrics as Ruled Surfaces

The equation of the hyperboloid of one sheet $\dfrac{x^2}{a^2} - \dfrac{z^2}{c^2} = 1 - \dfrac{y^2}{b^2}$ is satisfied by all values of x, y, z, which satisfy simultaneously the pair of equations

$$\frac{x}{a} - \frac{z}{c} = \mu\left(1 - \frac{y}{b}\right), \quad \frac{x}{a} + \frac{z}{c} = \frac{1}{\mu}\left(1 + \frac{y}{b}\right), \quad (1)$$

or the pair

$$\frac{x}{a} - \frac{z}{c} = \mu'\left(1 + \frac{y}{b}\right), \quad \frac{x}{a} + \frac{z}{c} = \frac{1}{\mu'}\left(1 - \frac{y}{b}\right), \quad (2)$$

when μ and μ' are parameters. For all values of μ equations (1) represent two planes whose intersection must lie on the hyperboloid. Likewise equations (2) for all values of μ' represent two planes whose intersection must lie on the hyperboloid. There are therefore two systems of straight lines generating the hyperboloid of one sheet.

Each straight line of one system is cut by every straight line of the other system. For the four equations (1) and (2) made simultaneous are equivalent to the three equations

$$\frac{x}{a} - \frac{z}{c} = \mu\left(1 - \frac{y}{b}\right), \quad \frac{x}{a} + \frac{z}{c} = \frac{1}{\mu}\left(1 + \frac{y}{b}\right), \quad \mu'\left(1 + \frac{y}{b}\right) = \mu\left(1 - \frac{y}{b}\right),$$

SECOND DEGREE EQUATION

from which
$$\frac{y}{b} = \frac{\mu - \mu'}{\mu + \mu'}, \quad \frac{x}{a} = \frac{1 + \mu\mu'}{\mu + \mu'}, \quad \frac{z}{c} = \frac{1 - \mu\mu'}{\mu + \mu'}.$$

No two straight lines of the same system intersect. Write the equations of lines of the first system corresponding to μ_1 and μ_2. Making the equations simultaneous $(\mu_1 - \mu_2)\left(1 - \frac{y}{b}\right) = 0$, and $\left(\frac{1}{\mu_1} - \frac{1}{\mu_2}\right)\left(1 + \frac{y}{b}\right) = 0$. Hence either $\mu_1 = \mu_2$, or $y = b$ and $y = -b$. Since y cannot be at once $+b$ and $-b$, $\mu_1 = \mu_2$; that is, two lines of the same system can intersect only if they coincide.

Observing that the equation of the hyperbolic paraboloid $\frac{z^2}{c} - \frac{y^2}{b} = 2x$ is satisfied by the values of x, y, z, which satisfy either of the pairs of equations

$$\frac{z}{\sqrt{c}} - \frac{y}{\sqrt{b}} = \frac{2x}{\mu}, \quad \frac{z}{\sqrt{c}} + \frac{y}{\sqrt{b}} = \mu, \tag{1}$$

$$\frac{z}{\sqrt{c}} + \frac{y}{\sqrt{b}} = \frac{2x}{\mu'}, \quad \frac{z}{\sqrt{c}} - \frac{y}{\sqrt{b}} = \mu', \tag{2}$$

it can be shown that this surface may be generated by two systems of straight lines; that each line of one system is intersected by every line of the other, and that no two lines of the same system intersect.

The equations of ellipsoid, hyperboloid of two sheets and of elliptical paraboloid cannot be resolved into real factors of the first degree, consequently these surfaces cannot be generated by systems of real straight lines.

Art. 112.—Asymptotic Surfaces

From the equation of the hyperboloid of one sheet
$$\frac{x^2}{a^2} + \frac{y^2}{b^2} - \frac{z^2}{c^2} = 1,$$

it is found that

$$\frac{z}{c} = \left(\frac{x^2}{a^2}+\frac{y^2}{b^2}\right)^{\frac{1}{2}}\left(1-\frac{a^2b^2}{a^2y^2+b^2x^2}\right)^{\frac{1}{2}} = \left(\frac{x^2}{a^2}+\frac{y^2}{b^2}\right)^{\frac{1}{2}} - \frac{ab}{2(a^2y^2+b^2x^2)^{\frac{1}{2}}}+\cdots,$$

the powers of $a^2y^2 + b^2x^2$ in the denominators increasing in the expansion by the binomial formula. Hence the z of the hyperboloid $\frac{x^2}{a^2}+\frac{y^2}{b^2}-\frac{z^2}{c^2}=1$, and the z of the cone

$$\frac{x^2}{a^2}+\frac{y^2}{b^2}-\frac{z^2}{c^2}=0$$

approach equality as x and y are indefinitely increased; that is, the conical surface is tangent to the hyperboloid at infinity. In like manner the cone $\frac{x^2}{a^2}-\frac{y^2}{b^2}-\frac{z^2}{c^2}=0$ is shown to be asymptotic to the hyperboloid of two sheets $\frac{x^2}{a^2}-\frac{y^2}{b^2}-\frac{z^2}{c^2}=1$.

ART. 113. — ORTHOGONAL SYSTEMS OF QUADRICS

The equation (1) $\frac{x^2}{a^2+\lambda}+\frac{y^2}{b^2+\lambda}+\frac{z^2}{c^2+\lambda}=1$, where $a>b>c$ and λ is a parameter, represents an ellipsoid when $\infty > \lambda > -c^2$, an hyperboloid of one sheet when $-c^2 > \lambda > -b^2$, an hyperboloid of two sheets when $-b^2 > \lambda > -a^2$, an imaginary surface when $\lambda < -a^2$.

Through every point of space (x', y', z') there passes one ellipsoid, one hyperboloid of one sheet, and one hyperboloid of two sheets of the system of quadrics represented by equation (1). For, if λ is supposed to vary continuously from $+\infty$ to $-\infty$ through 0, the function of λ,

$$\frac{x'^2}{a^2+\lambda}+\frac{y'^2}{b^2+\lambda}+\frac{z'^2}{c^2+\lambda}-1,$$

is — when $\lambda = +\infty$ and $+$ when λ is just greater than $-c^2$, — when λ is just less than $-c^2$ and $+$ when λ is just greater than $-b^2$, — when λ is just less than $-b^2$ and again $+$ when λ is just greater than $-a^2$. Hence

$$\frac{x'^2}{a^2+\lambda} + \frac{y'^2}{b^2+\lambda} + \frac{z'^2}{c^2+\lambda} - 1 = 0 \qquad (2)$$

must determine three real values for λ; one between $+\infty$ and $-c^2$, another between $-c^2$ and $-b^2$, a third between $-b^2$ and $-a^2$.

Let $\lambda_1, \lambda_2, \lambda_3$ be the roots of equation (2); that is, let

$$\frac{x'^2}{a^2+\lambda_1} + \frac{y'^2}{b^2+\lambda_1} + \frac{z'^2}{c^2+\lambda_1} = 1, \qquad (3)$$

$$\frac{x'^2}{a^2+\lambda_2} + \frac{y'^2}{b^2+\lambda_2} + \frac{z'^2}{c^2+\lambda_2} = 1, \qquad (4)$$

$$\frac{x'^2}{a^2+\lambda_3} + \frac{y'^2}{b^2+\lambda_3} + \frac{z'^2}{c^2+\lambda_3} = 1. \qquad (5)$$

The equations of tangent planes to the quadrics of system (1) corresponding to $\lambda_1, \lambda_2, \lambda_3$ at the point of intersection (x', y', z') are

$$\frac{xx'}{a^2+\lambda_1} + \frac{yy'}{b^2+\lambda_1} + \frac{zz'}{c^2+\lambda_1} = 1,$$

$$\frac{xx'}{a^2+\lambda_2} + \frac{yy'}{b^2+\lambda_2} + \frac{zz'}{c^2+\lambda_2} = 1,$$

$$\frac{xx'}{a^2+\lambda_3} + \frac{yy'}{b^2+\lambda_3} + \frac{zz'}{c^2+\lambda_3} = 1.$$

The condition of perpendicularity of the first two planes

$$\frac{x'^2}{(a^2+\lambda_1)(a^2+\lambda_2)} + \frac{y'^2}{(b^2+\lambda_1)(b^2+\lambda_2)} + \frac{z'^2}{(c^2+\lambda_1)(c^2+\lambda_2)} = 0$$

is a consequence of (3) and (4). In like manner it is shown that the three tangent planes are mutually perpendicular.

Hence equation (1) represents an orthogonal system of quadrics.

Since through every point of space there passes one ellipsoid, one hyperboloid of one sheet, and one hyperboloid of two sheets of the orthogonal system of quadrics, the point in space is determined by specifying the quadrics of the orthogonal system on which the point lies. This leads to elliptic coordinates in space, developed by Jacobi and Lamé in 1839, by Jacobi for use in geometry, by Lamé for use in the theory of heat.

If a bar kept at a constant temperature is placed in a homogeneous medium, when the heat conditions of the medium have become permanent the isothermal surfaces are the ellipsoids, the surfaces along which the heat flows the hyperboloids, of the orthogonal system of quadrics.

www.ingramcontent.com/pod-product-compliance
Lightning Source LLC
Chambersburg PA
CBHW021832230426
43669CB00008B/941